VIII International Conference on Nanomaterials and Technologies 2019

IOP Conference Series: Materials Science and Engineering
Volume 704

Ulan-Ude, Russia
23 - 28 August 2019

ISBN: 978-1-7138-1034-6
ISSN: 1757-8981

Printed from e-media with permission by:

Curran Associates, Inc.
57 Morehouse Lane
Red Hook, NY 12571

Some format issues inherent in the e-media version may also appear in this print version.

Institute of Physics
Dirac House, Temple Back
Bristol BS1 6BE UK

Phone: 44 1 17 929 7481
Fax: 44 1 17 920 0979

techtracking@iop.org

Additional copies of this publication are available from:

Curran Associates, Inc.
57 Morehouse Lane
Red Hook, NY 12571 USA
Phone: 845-758-0400
Fax: 845-758-2633
Email: curran@proceedings.com
Web: www.proceedings.com

TABLE OF CONTENTS

IOP Conf. Series: Materials Science and Engineering **704** (2019) 011001 doi:10.1088/1757-899X/704/1/011001

This collection contains reports of the VIII International Conference "Nanomaterials and Technologies" (August 24–28, 2019, Ulan-Ude - Lake Baikal).

The collection includes reports of scientists from famous scientific centers. The scientific program of the conference included questions of the formation, identification of the mechanisms of formation and characterization of various nanoscale objects using electron beams, laser irradiation, the sol-gel method, in an electric discharge plasma in a gas. The reports present studies of the structure and properties of nanoparticles, nanotubes, and other nano-objects created by the mentioned methods.

As a result of modification of substances and materials by nanoobjects, their properties, physical and chemical characteristics change: mechanical, magnetic, electrophysical, catalytic, optical, viscosity. Another important area of study of nanoscale objects is their use in various compounds, materials, such as steel, liquid crystals, motor oils, microcapillary reactors, fuel cells in order to modify and improve their properties.

Research in the field of nanomaterials is an interdisciplinary science, primarily based on physical and chemical methods. Therefore, the works presented in the collection in the field of physics and chemistry of the condensed state, describing the behavior of glassy systems, liquid materials, and the use of X-ray diffraction methods, become logical.

The mathematical and computer modeling of materials presented in the reports is effective methods for studying complex systems, such as nano-objects, consisting of a limited number of atoms and molecules subject to computer calculation, which makes it possible not only to model possible stable structures, but also depending on their size, physical parameters - temperature, pressure and others.

Program Committee:

- Prof. **Sergey Bardakhanov**, Co-Chairman (Institute of theoretical and applied mechanics SB RAS)
- Prof. **Andrey Nomoev**, Co-Chairman (Institute of Physical Material Science SB RAS)
- Prof.Dr. **Jav Davaasambuu**, Co-Chairman (Mongolian Physical Socie-ty)
- Prof. **Ojin Tegus,** Co-Chairman (Inner Mongolia Normal University),
- Member of RAS **A.L. Aseev**, Honored Chairman (Institute of Semicon-ductor Physics SB RAS, Novosibirsk).
- Member of RAS **L.A. Smirnov** (Ural State Technical University),
- Member of RAS **V.M. Schastlivtsev** (Institute of Metal Physics UB RAS),
- Member of RAS **Ju.A. Zolotov** (Kurnakov Institute of General and In-organic Chemistry),
- Member of RAS **L.B. Boinovich** (Institute of Physical Chemistry and Electrochemistry RAS),
- Member of RAS **L.I. Leontiev (**Institute of Metallurgy UrB RAS),
- Member of RAS **V.L. Kozhevnikov**(Institute of Solid State Chemistry UB AS),
- Member of RAS **V.P. Meshalkin** (University of Chemical Technology of Russia),
- RAS Corresponding Member **A.G. Kolmakov** (Institute of Metallurgy and Material Science RAS)
- RAS Corresponding Member **I.V. Melikhov (**Chemical Department of Moscow State University)**,**
- RAS Corresponding Member **A.B. Yaroslavtsev** (Kurnakov Institute of General and Inorganic Chemistry),
- RAS Corresponding Member **V.F. Balakirev** (Institute of Metallurgy UB RAS),
- RAS Corresponding Member **V. V. Sagaradze** (M.N. Mikheev Institute of Metal Physics UB RAS),
- RAS Corresponding Member **S.V. Gnedenkov** (Institute of Chemistry FEB RAS),
- RAS Corresponding Member **S.Yu. Bratskaya**. (Institute of Chemistry FEB RAS),
- RAS Corresponding Member **Yu. G. Gorbunova** (Institute of Physical Chemistry and Electrochemistry RAS),
- RAS Corresponding Member **E. A. Gudilin** (Faculty of Material Sci-ence of Moscow State University),
- RAS Corresponding Member **V. L. Stolyarova** (St. Petersburg State University),

IOP Conf. Series: Materials Science and Engineering **704** (2019) 011002 doi:10.1088/1757-899X/704/1/011002

- RAS Corresponding Member **V. V. Gusarov** (St. Petersburg State In-stitute of Technology),
- RAS Corresponding Member **M. I. Alymov (**Institute of structural macrokinetics and Material Science RAS),
- RAS Corresponding Member **A. I. Nikolaev** (Institute of Chemistry and Technology of Rare Elements and Mineral Raw Materials RAS),
- RAS Corresponding Member **B. V. Gusev** (RUT-MIIT),
- RAS Corresponding Member **V. G. Bamburov** (Institute of Solid State Chemistry UB RAS),
- Acad. **Jadambaa Temujin** (Institute of Chemistry and Chemical Technology, MAS),
- Acad. **Deleg Sangaa** (Institute of Physics and Technology MAS),
- Prof. **Aleksandr Semenov** (Institute of Physical Material Science of SB RAS),
- Prof. **Lubov Bokhoeva** (East-Siberian State University of Technology and Management),
- Prof. **Dandar Sanditov** (Buryat State University),
- Prof. **Pagvajav Altantsog** (Institute of Physics and Technology, MAS),
- Acad. **O. Penyazkov** (A.V.Lykov Institute of Heat and Mass Transfer, Belarus),
- Prof. **G. Shilagardi** (National University of Mongolia),
- Prof. **Altan Bolag** (Inner Mongolia Normal University),
- Prof. **U. Pietsch** (University of Siegen, Germany).

Organizing Committee:

Prof. Andrey Nomoev, Co-Chairman (Institute of Physical Material Science of SB RAS)

Prof. Nikolay Moshkin, Co-Chairman (Buryat State University)

Dr. Vyacheslav Syzrantsev, Scientific Secretary (Institute of Physical Material Science of SB RAS)

Erzena Khartaeva, Secretary (Institute of Physical Material Science of SB RAS)

Prof. Lyubov Bokhoeva — Chairman of the organizing committee of the youth conference (East-Siberian State University of Technology and Management)

Prof. Vyacheslav Khakhinov (Buryat State University)

Prof. Igor Sizov (East Siberia State University of Technology and Management),

Prof. Shulun Tsydypov (Buryat State University)

Dr. Bair Damdinov (Siberian Federal University, Buryat State Univer-sity)

Dr. Tumen Darmaev (Buryat State University)

Valery Lygdenov (Buryat State University)

Sergey Kalashnikov (Buryat State University)

IOP Conf. Series: Materials Science and Engineering **704** (2019) 011002 doi:10.1088/1757-899X/704/1/011002

Bair Khabituev (Buryat State University)
Dr. N.Tovjargal (National University of Mongolia)
Dr. Tsetsenbaatar (Inner Mongolia Normal University)
M. Batpurev (National University of Mongolia)

Peer review statement

All papers published in this volume of *IOP Conference Series: Materials Science and Engineering* have been peer reviewed through processes administered by the proceedings Editors. Reviews were conducted by expert referees to the professional and scientific standards expected of a proceedings journal published by IOP Publishing.

IOP Conf. Series: Materials Science and Engineering **704** (2019) 012001 doi:10.1088/1757-899X/704/1/012001

Polyurethane foam modified with borohydride: preparation and possibilities of using for synthesis of gold nanoparticles and their nanocomposites

A I Isachenko[1], V V Apyari[1], A O Melekhin[1], A V Garshev[1,2], P A Volkov[3], S G Dmitrienko[1]
[1]Department of Chemistry, Lomonosov Moscow State University, Moscow, 119991 Russia
[2]Department of Materials Science, Lomonosov Moscow State University, Moscow, 119991 Russia
[3]Scientific-Research Institute of Chemical Reagents and Special Purity Chemicals of National Research Center "Kurchatov Institute", Bogorodsky Val, 3, Moscow 107076, Russia

E-mail: isandrey91@gmail.com

Abstract. The present work describes preparation of polyurethane foam modified with borohydride. Adsorption of borohydride from alkaline solutions in presence of a cationic surfactant is discussed. The polyurethane foam modified with borohydride possesses reductive ability with respect to Au(III). Possibilities of the new material for synthesis of gold nanoparticles are described. This solid reducting agent is convenient for use in the synthesis of nanoparticles, since it ensures precise dosing of a reductant while desorbed from polymer. Gold nanoparticles can be synthesized both in solution and on the polymer surface by varying reagents concentrations during the preparation of modified polyurethane foam.

1. Introduction

Gold nanoparticles possess a number of valuable qualities: unique optical properties due to surface plasmon resonance (SPR), highly developed surface, low toxicity, chemical inertness, catalytic activity, etc [1-6].

The method and conditions of gold nanoparticles preparation in many cases determines their properties: morphological, optical, and other features [1-5, 7]. Often, nanoparticles functionalization occurs during their preparation, which determines selectivity and sensitivity of the synthesized reagent. Modern research is aimed at obtaining gold nanoparticles with various sizes and a narrow size distribution.

Spherical gold nanoparticles are most often obtained by chemical reduction of hydrochloric acid with sodium borohydride. The main feature of sodium borohydride is its great reducing ability. Sodium borohydride is quite stable in solid form and convenient for use in the synthesis of nanoparticles in both aqueous and organic media.

The size and dispersion of the resulting nanoparticles, as well as their stability over time, are controlled by varying the nature of a stabilizer and its amount. As stabilizers, an excess of a reducing agent is used, as well as specially introduced substances. Often for the synthesis of monodispersed gold nanoparticles, ionic surfactants are used, for example sodium dodecyl sulfate [8] or cetyltrimethylammonium bromide [9], as well as synthetic and natural polymers — polyvinylpyrrolidone [10], polyethylene glycol [11], cyclodextrins [12], polyamidoamine [13] and etc.

To stabilize and improve the operational performance of gold nanoparticles, they can be immobilized on various solid matrices. The nanocomposites obtained in this way can advantageously differ in their optical and analytical characteristics. A large number of different matrices used to obtain nanocomposite materials is described. Among the variety of matrices, polyurethane foams (PUFs) should be considered specifically as having a number of quite important advantages: good sorption ability, low density (0.015 – 0.045 g cm^{-3}), monolithicity, resistance to thermo-oxidative degradation, chemical inertness to many compounds, low cost and the availability of these materials [14-15].

In this work the features of polyurethane foam modification with borohydride for subsequent use as a reducing agent in the preparation of gold nanoparticles and their nanocomposites is studied and discussed.

2. Experimental

2.1. Materials

The following reagents at least of analytical grade were used: sodium borohydride, cetyltrimethilammonium bromide (CTAB), sodium hydroxide, hydrogen tetrachloroaurate, hydrochloric acid. Stock solutions of these reagents were prepared by dissolving their weighed portions or aliquots in deionized water.

Polyether-based polyurethane foam (PUF) was used. The polymer was cut into cylindrical tablets of 16 mm in diameter. The weight of each tablet was (20 ± 2) mg. To clean samples, the PUF tablets were placed in acetone and shaken for 10 minutes. The procedure was repeated twice, and then the tablets were dried under stream of air. The tablets were stored in a light-protected place.

2.2. Instruments

Absorption spectra of solutions were recorded using SF-103 spectrophotometer (Akvilon, Russia), diffuse reflectance measurements were carried out using an Eye-One Pro mini-spectrophotometer (X-Rite) [16, 17] on a white base. The diffuse reflectance coefficients (R) were recalculated into the Kubelka-Munk function (F) values to get absorption spectra in a solid phase according to the formula: $F = \frac{(1-R)^2}{2R}$. pH was measured using Ekspert 001 pH-meter (Ekoniks Ekspert). TEM-images were recorded using transmission electron microscope Libra 200 (Zeiss, Germany) at the accelerating voltage of 200 kV. Aqueous dispersions of the samples were deposited onto copper grid support with a formvar film covered by amorphous carbon Formvar®/Carbon Reinforced Copper Grids 3440C-MB (SPI, USA). Scanning electron microscopic studies of PUF samples microstructure was carried out using scanning electron microscope JSM 7100 F (Jeol, Japan) at the accelerating voltage of 2–5 kV. Deionized water was obtained using the Millipore Simplicity purification system (Millipore).

2.3. Adsorption of borohydride onto polyurethane foam

Adsorption of borohydride onto PUF was carried out in a static mode. The PUF tablet was placed into a test-tube containing 5 mL of an following mixture: 0.01 mol L^{-1} sodium hydroxide, 0.01 mol L^{-1} sodium borohydride and 0.0001 mol L^{-1} CTAB for further application of the modified polymer in the synthesis of nanocomposites or 0.1 mol L^{-1} sodium borohydride and CTAB for further application of the modified polymer in the synthesis of gold nanoparticles colloids. The tablets were thoroughly pressed with a glass rod to remove air bubbles from pores and shaken on an electromechanical shaker for some time. After that, the tablets were removed and dried between sheets of filter paper.

To control adsorption of borohydride, freshly prepared modified tablets were placed in 5 mL of hydrogen tetrachloroaurate containing 20 μg mL^{-1} of gold. The tablets were thoroughly pressed with a glass rod and stirred by shaking on an electromechanical shaker for 30 minutes. Then, the tablets were removed and dried between sheets of filter paper

3. Results and discussion

In this work, PUF has been first modified with borohydride aiming for its following use as a new reductant for synthesis of gold nanoparticles colloids and nanocomposites.

Sorption of sodium borohydride onto PUF was carried out from a solution containing sodium hydroxide and CTAB. Sodium borohydride is a polar compound that dissociates well into ions in aqueous solutions. In order to overcome the interfacial boundary and ensure transition of borohydride anions to the PUF phase, CTMA was added to the solution as a hydrophobic counterion. Since in aqueous solutions, especially in an acidic environment, borohydride is unstable, NaOH was added to the system as a stabilizer.

Adsorption of borohydride onto PUF in the mentioned conditions was studied in detail. Sorption of sodium borohydride onto PUF was carried out from a solution containing sodium hydroxide and CTAB. Sodium borohydride is a polar compound that dissociates well into ions in aqueous solutions. In order to overcome the interfacial boundary and ensure transition of borohydride anions to the PUF phase, CTMA was added to the solution as a hydrophobic counterion. Since in aqueous solutions, especially in an acidic environment, borohydride is unstable, NaOH was added to the system as a stabilizer.

Adsorption of borohydride onto PUF in the mentioned conditions was studied in detail.

3.1. Adsorption kinetics of borohydride on polyurethane foam

For kinetics study of borohydride adsorption by polyurethane foam, a number of experiments were carried out according to the scheme described above with different mixing times at the first stage at $NaBH_4$ concentration of 0.01 M. Diffuse reflectance spectra for borohydride-modified tablets treated with a solution of hydrogen tetrachloroaurate were recorded (figure 1a). The dependence of the Kubelka-Munk function obtained at the absorption maximum at 540 nm on the time of adsorption of borohydride is shown in figure 1b.

According to the graphics, 15 minutes is enough to achieve a maximum of the function. This time was used in further experiments. A slight decrease at times greater than 20 min can be explained by the gradual decomposition of sodium borohydride with subsequent stirring.

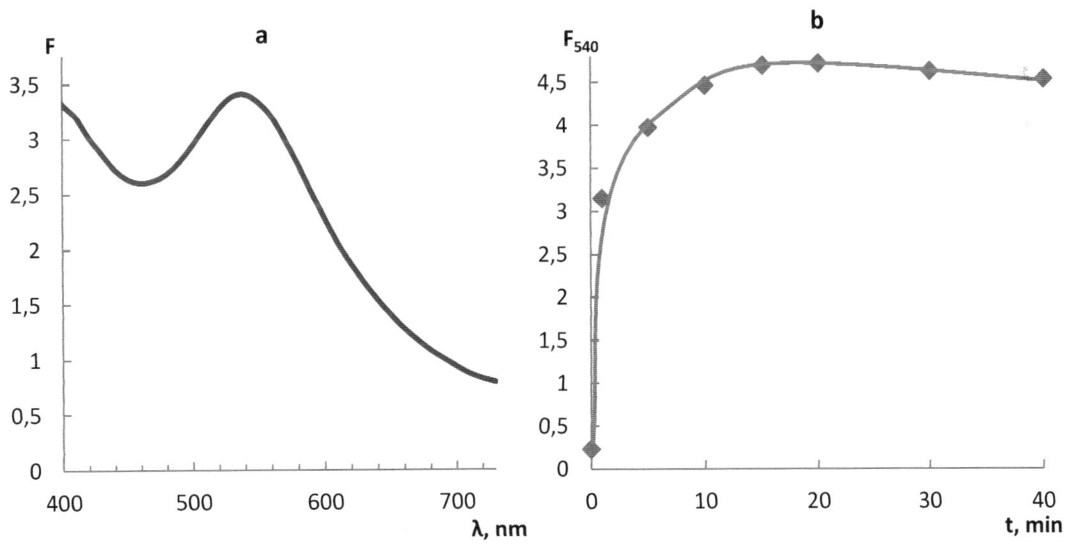

Figure 1. a) Diffuse reflectance spectrum of PUF modified with gold nanoparticles; b) dependence of the Kubelka-Munk function at 540 nm on the phases contact time of the PUF with a borohydride solution.

For numerical evaluation of adsorption kinetics of sodium borohydride on polyurethane foam several classical kinetic models of adsorption were used: the Lagergren pseudo first order model

($\ln(1 - a/a_{eq}) = -k_1 t$) and the pseudo second order model ($t/a = t/a_{eq} + 1/k_2 \cdot a^2_{eq}$), where k_1 is the pseudo first order reaction rate constant, min^{-1}; k_2 is the pseudo-second order reaction rate constant, $g \cdot \mu mol^{-1} \cdot min^{-1}$); a_{eq} — adsorption value upon reaching the sorption equilibrium between the sorbent and sorbate, $\mu mol \cdot g^{-1}$; t is the adsorption time, min.

In order to ascertain the sorption mechanism and select the kinetic model that most adequately describes the sorption of borohydride on polyurethane foam, the integral kinetic curve was processed and linearized taking into account the fact that the Kubelka-Munk function is directly proportional to the specific adsorption. The obtained dependences are presented in figure 2.

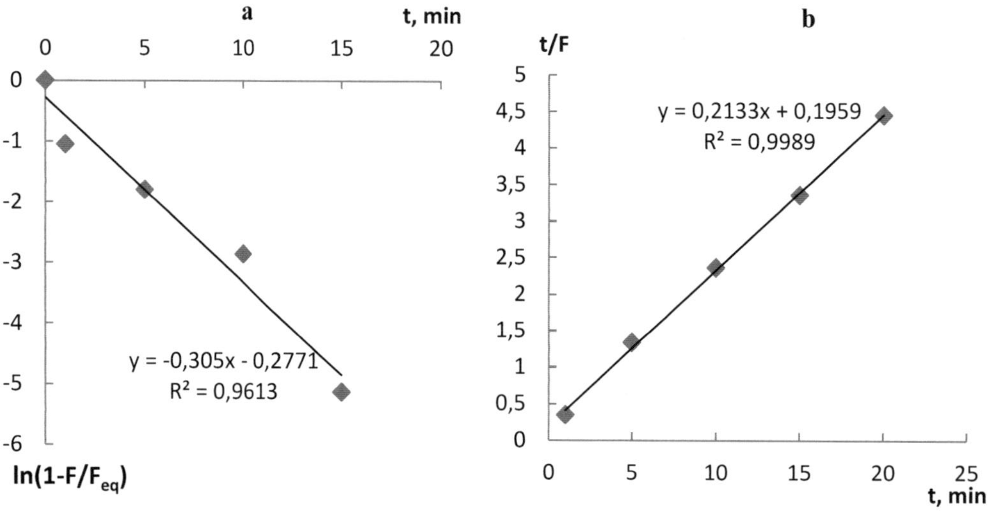

Figure 2. Kinetic dependences of the adsorption of sodium borohydride on polyurethane foam linearized using the pseudo-first (a) and pseudo-second (b) order models.

According to the obtained graphics, the adsorption kinetics is best described by the pseudo-second order model ($R^2 = 0.999$), $F_{eq} = 4.69$. The calculated value of the pseudo-second order adsorption rate constant was 0.232 min^{-1}. The proportionality coefficient between the Kubelka-Munk function and the specific adsorption of borohydride is 18385 $g \cdot mol^{-1}$, therefore the pseudo-second order adsorption rate constant can be calculated as $4.3 \cdot 10^{-3}$ $g \cdot \mu mol^{-1} \cdot min^{-1}$.

3.2. Adsorption isotherm.

In order to construct the adsorption isotherm of borohydride on polyurethane foam, a number of experiments were carried out according to the scheme described above, varying the concentration of sodium borohydride in the range of 0.001 — 0.02 M.

After removing PUF tablet, the solution was neutralized with an excess of hydrochloric acid of known concentration. Hydrochloric and boric acids in the joint presence were determined using potentiometric titration with alkali with a glass pH-selective electrode. First, hydrochloric acid was titrated until a sharp change in pH. Since boric acid cannot be titrated directly by acid-base titration due to the low acidity constant (pK_a = 9.24), glycerol was added to the system during its determination. After this, titration was continued until a second pH increase, which corresponded to boric acid. From the obtained data, equilibrium concentration of borohydride in the aqueous solution was calculated. The specific adsorption of the reducing agent was found by the difference between the initial and equilibrium concentrations divided by the mass of polyurethane foam. The obtained adsorption isotherm of borohydride is shown in figure 3.

It has two stages. It was determined that the first step can be described by the Langmuir adsorption isotherm with parameters K = 724 L mol^{-1}, $a_{max} = 3.1 \cdot 10^{-4}$ mol g^{-1}. Presumably, it is related to the adsorption of borohydride on the surface of polyurethane foam and its absorption near the surface of polymer membranes. The second stage — more pronounced — can be caused by the absorption of borohydride in the polymer volume, which is often observed when studying the sorption of various organic substances on polyurethane foam [14]. This is also confirmed by the significant difference between the experimental and theoretically calculated values for the maximum adsorption of borohydride, which amounted to $8.0 \cdot 10^{-6}$ mol g^{-1} ($a_m = s_{sp} / (s_{mol}N_A)$, the specific surface of the polyurethane foam is 0.3 m^2 g^{-1} [14], and the size of the BH_4^- ion is 2.5 Å).

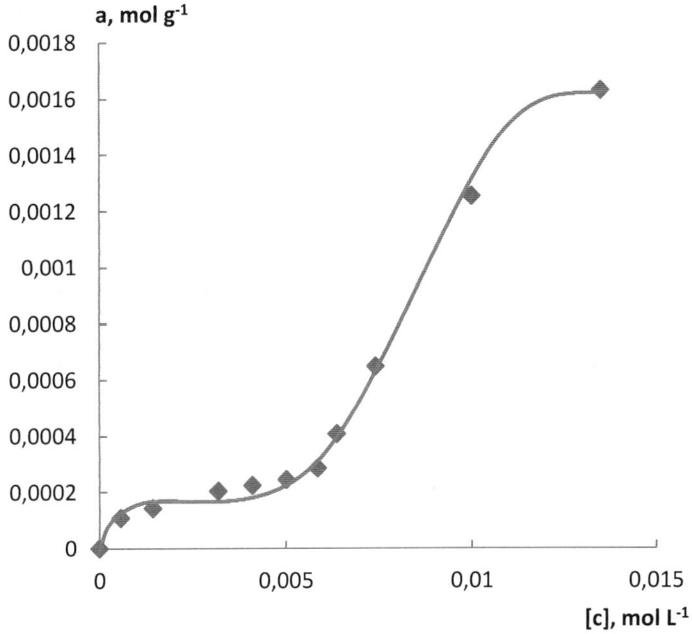

Fig. 3. Adsorption isotherm of borohydride on polyurethane foam.

3.3. Synthesis of gold nanoparticles and their nanocomposites

We found that polyurethane foam modified with borohydride reduces Au(III) from a solution to form gold nanoparticles on PUF surface. The resulting material is a polymer-based nanocomposite with a characteristic SPR band in the spectrum located at 540 nm.

The supposed mechanism of the nanocomposite formation includes slow diffusion-controlled desorption of borohydride anions from polyurethane foam and their interaction in the surface layer with Au(III) to form gold nanoparticles. In absence of a stabilizer that facilitates gold nanoparticles transition to a colloidal solution, they attach onto the surface. In this case, PUF plays a role of a solid-phase stabilizer, forming the nanocomposite.

In this view, it is important to discuss the role of CTAB. On the one hand, this substance promotes sorption of borohydride anions onto PUF, forming a hydrophobic ionic associate. On the other hand, CTAB is a stabilizer of nanoparticles [18] that promote formation of a colloidal system in a solution at high concentration. Depending on the concentration of CTAB, the higher amount of gold nanoparticles on PUF, the lower their amount in the solution can be obtained, and vice versa. At CTAB concentration of 0 and 0.1 M, gold nanoparticles preferably exist in the aqueous solution, at 0.0001 – 0.001 M CTAB they form the nanocomposite. It shows the possibility of controlling distribution of nanoparticles between phases. Increasing $NaBH_4$ concentration up to 0.1 M leads to a sharper dependence of the nanocomposite absorbance on concentration of CTAB.

Nanocomposites of gold were studied using scanning electron microdcopy. Formed nanoparticles are attached on a polymer network. A SEM image (Fig. 4a) shows that they are evenly distributed over the surface of the polymer. Nanoparticles on PUF surface have mainly spherical shape. Their average diameter is 12 ± 5.

Colloids of gold nanopaticles by increasing the concentration of CTAB and $NaBH_4$ during preparation of borohydride-modified PUF were obtained. Fig. 4b shows a TEM image of gold nanopaticles synthesized. Their average diameter is (4 ± 1) nm. They have narrow size distribution, which can be ascribed to the slow release of borohydride during desorption from PUF avoiding high local supersaturation and formation of many small particles. On the other hand, this can be ascribed to the stabilizing action of the polymer matrix with respect to nanoparticles forming near its surface.

a b

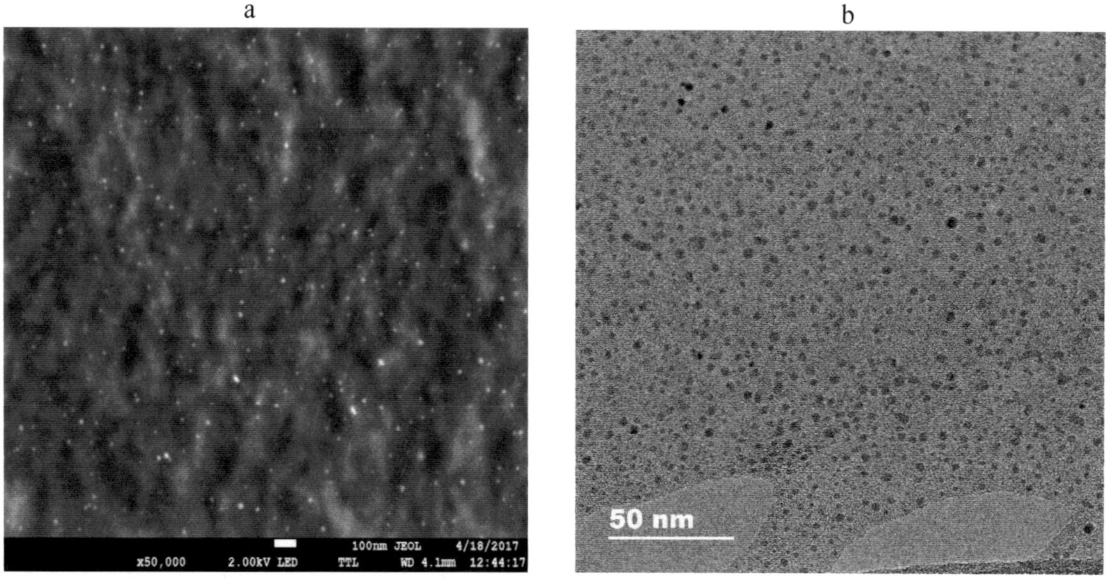

Figure 4. *A SEM image of gold nanocomposites with PUF (a) and TEM image of gold nanoparticles in solution (b).*

4. Conclusions

The features of borohydride-modified polyurethane foam synthesis were studied. This solid-phase reagent can be obtained by a simple adsorption technique. The kinetics of adsorption of sodium borohydride on polyurethane foam was studied. The kinetic sorption curve is well described by a pseudo-second order model. The adsorption isotherm of sodium borohydride on polyurethane foam consists of two stages indicating complex mechanism of adsorption. First stage can be described by the Langmuir equation.

Polyurethane foam modified with borohydride was found to reduce Au(III) from solution to form nanocomposites or colloidal solutions of gold nanoparticles. Due to the controlled desorption of borohydride from polyurethane foam, the synthesized nanoparticles has a narrow size distribution. Control of nanoparticles distribution between polymer and solution can be easily performed by varying CTAB and $NaBH_4$ concentration during the adsorption. The presence of optical properties in the nanocomposites allows their possible use as sensitive materials for optical sensors.

Acknowledgments

This work was financially supported by the Russian Science Foundation (grant N 18-73-10001). Some studies were performed using instrumentation provided according to the M.V. Lomonosov Moscow State University Program of Development

References

[1] Dykman L A, Bogatyrev V A, Shchegolev S Yu, Khlebtsov N G 2008 *Gold Nanoparticles: Synthesis, Properties and Biomedical Use* (Moscow: Nauka).

[2] Patra C R, Bhattacharya R, Mukhopadhyay D, Mukherjee P. 2010 Fabrication of gold nanoparticles for targeted therapy in pancreatic cancer *Adv. Drug. Deliver. Rev.*. **62**. 346-361

[3] Ghosh S K, Pal T 2007 Interparticle Coupling effect on the surface plasmon resonance of gold nanoparticles: from theory to applications *Chem. Rev.* **107** 4797–4862
[4] Saarinen J J, Vartiainen E M, Peiponen K-E 2009 Surface plasmon resonance reflectance from nanoparticles in a liquid matrix: Retrieval of the optical properties using the maximum entropy model *Sensor. Actuat. B* **138** 383–395
[5] Uechi I, Yamada S 2008 Photochemical and analytical applications of gold nanoparticles and nanorods utilizing surface plasmon resonance. *Anal. Bioanal. Chem.* **391** 2411–2421
[6] Daniel M-C, Astruc D. 2004 Gold nanoparticles: assembly, supramolecular chemistry, quantum-size related properties, and applications towards biology, catalysis and nanotechnology *Chem. Rev.* **104** 293–346
[7] Nguyen D T, Kim D-J, Kim K-S 2011 Controlled synthesis and biomolecular probe application of gold nanoparticles *Micron.* **42** 207–227
[8] Lakade A J, Sundar K, Shetty P H 2018 Gold nanoparticle-based method for detection of calcium carbide in artificially ripened mangoes. *Food Addit. Contam.* **35** 1078–1084
[9] Adegoke O, McKenzie C, Daeid N N 2019 Multi-shaped cationic gold nanoparticle-L-cysteine-ZnSeS quantum dots hybrid nanozyme as an intrinsic peroxidase mimic for the rapid colorimetric detection of cocaine *Sens. Actuat. B* **287** 416–427
[10] Mohamed T, Matou-Nasri S, Farooq A, Whitehead D, Azzawi M 2017 Polyvinylpyrrolidone-coated gold nanoparticles inhibit endothelial cell viability, proliferation, and ERK1/2 phosphorylation and reduce the magnitude of endothelial-independent dilator responses in isolated aortic vessels. *Int. J. Nanomedicine.* **12** 8813–8830
[11] Chen X-Y, Ha W, Shi Y-P 2019 Sensitive colorimetric detection of melamine in processed raw milk using asymmetrically PEGylated gold nanoparticles *Talanta* **194** 475–484
[12] Dykman L A, Bogatyrev V A 2007 Gold nanoparticles: preparation, functionalization and applications in biochemistry and immunochemistry *Usp. Khim.* **76** 199–213
[13] Guo Y, Liang X, Bi J, Ling R, Jiang Y, Mou Z, Huang F, Qin W 2019 A polyamidoamine-mediated competitive colorimetric assay based on gold nanoparticles for determining acid values in edible sunflower seed, corn and extra virgin olive oils *Food Chem.* **285** 450–457
[14] Dmitrienko S G, Apyari V V 2010 *Penopolyurethany: sorbtsionnye svoistva i primenenie v khimicheskom analize (Polyurethane foams: sorption properties and application in chemical analysis)* (Moskow: Krasand) p 264
[15] Panigrahi S, Kundu S, Ghosh S K, Nath S, Pal T 2004 General method of synthesis for metal nanoparticles *J. Nanopart. Res.* **6** № 4 411–414
[16] Apyari V V; Dmitrienko S G, Zolotov Yu A 2013 Unusual application of common digital devices: potentialities of Eye-One Pro mini-spectrophotometer – a monitor calibrator for registration of surface plasmon resonance bands of silver and gold nanoparticles in solid matrices *Sensors Actuat. B* **188** 1109–1115
[17] Apyari V V, Dmitrienko S G, Batov I V, Zolotov Yu A 2011 An Eye-One Pro mini-spectrophotometer as an alternative to diffuse reflectance spectrometer *J. Anal. Chem.* **66** 144–150
[18] Tan K S, Cheong K Y 2013 Advances of Ag, Cu, and Ag–Cu alloy nanoparticles synthesized via chemical reduction route *J. Nanopart. Res.* 15:1537 DOI: 10.1007/s11051-013-1537-1

IOP Conf. Series: Materials Science and Engineering **704** (2019) 012002 doi:10.1088/1757-899X/704/1/012002

Modelling uranyl-ions sorption by nanostructured composite anatase-based materials

T A Sokolnitskaya[1], V V Zheleznov[1], A P Golikov[1] and A E Plakhutina[2]

[1] Institute of Chemistry FEB RAS, Prospect 100-letiya Vladivostoka 159. Vladivostok, 690022, Russia

[2] – Far Eastern Federal University, Sukhanova str. 8. Vladivostok, 690050, Russia

E-mail: ttt@ich.dvo.ru

Abstract. Influence of complex formation in solution on U sorption has been studied for mesoporous nanostructured anatase-based composites. Uranium sorption degree has been shown to correlate with the stability of the complexes in solution. A model for uranium sorption by titania materials has been suggested presuming two-stage sorption mechanism. It is assumed that initial binding of neutral and positively charged uranyl complexes to negatively charged hydroxyl groups occurs on the first stage, while decomposition of complex ion proceeds on the second stage with the UO_2 being fixed on sorption sites. Prevailing uranium species in solution and kinetics of their ratios have been determined. Calculated rates of evolution of uranyl ion species distribution (therefore, rate and efficiency of the sorption) correlate well with complex stability in solution.

1. Introduction

Intensive mining of Uranium ores produces large amounts of industrial waste such as tailing ponds and ore dumps. As a result of weathering processes Uranium compounds can migrate at large distances which poses significant ecological hazards due their high toxicity. MPC of Uranium in drinking water is 15mkg/L [1], and groundwater concentration can achieve 40mkg/L in contaminated areas [2].

Development of new modern sorption materials tailored for Uranium concentration and recovering is the important area of applied science focused both on further development of industry producing the products containing Uranium and on cleanup of contaminated environment from uranium compounds.

Recent studies show that increasing of sorption capacity and improvement of kinetic characteristics of inorganic compounds can be achieved by regulation of pore structure and increasing the surface area; uptake selectivity can be significantly boosted by using materials with high affinity to Uranil-Ions such as titanium silicates and zirconium silicates.

Sorption of Uranium from low concentration model solutions containing sulphate, fluoride or oxalate ions onto mesoporous nanostructured materials of Titanium oxide doped by Zirconium Oxide has been studied.

2. Results

Approach based on usage of the rigid latex particles with low polydispersity index as templates and meticulous selection of template multy step burning out conditions allows to produce materials with controlled pore structure and developed surface (table 1).

Table 1. Composition and structure of several samples

Sample	Composition, %			Specific surface area, m^2/g	Specific pore volume, sm^3/g	Phase composition
	TiO_2	ZrO_2	SiO_2			
TЭ0	95,0	0,0	5,0	155	0,45	anatase
TЭ5	88,4	5,6	6,0	170	0,49	anatase

Factors which affect Uranyl -ions sorption:

- Amount and characteristics of active centers on the sorbent surface (it is mostly defined by type and concentration of dope).

It was shown that materials composed of nanosize TiO_2 have surficial hydroxo complexes due to the surficial and volumetric oxygen vacancies and demonstrating very high reactivity due to excess electron [5]. It is the hydroxo complexes that define sorption activity of TiO_2.

- Degree of protonation of active centers (sorption activity and pH of solution dependency).

As pH rises above 6-7 the Uranium recovery decline was observed. For neutral pH the degree of recovery approaches 100% after only 2 hours of sorption (Figure 1). As Titanium Oxide has amphoteric properties the sorption from acidic solutions pH slightly increases and in weak alkaline systems pH decreases. Similar dependencies were observed in all three studied systems.

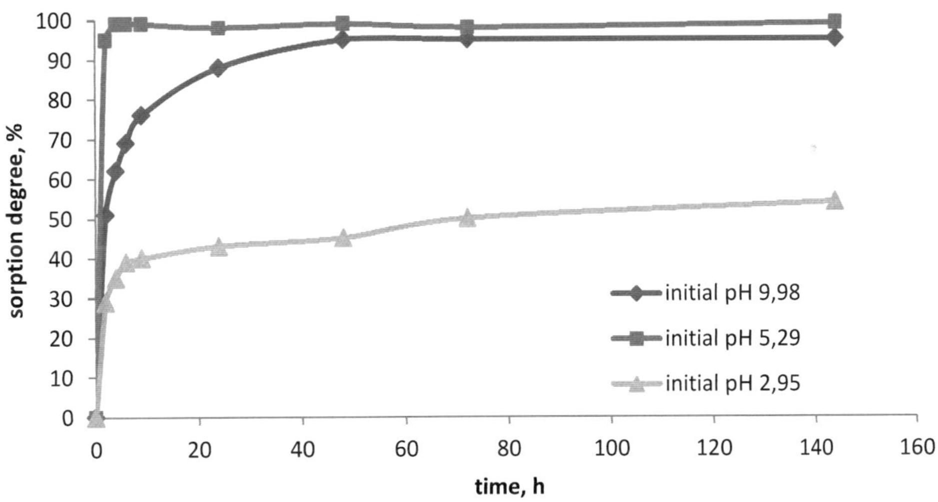

Figure 1. Sorption kinetics of U (VI). (Sulfate system. $C(U) = 4{,}2\cdot10^{-5}$ mol/l, $C(SO_4^{-2}) = 5\cdot10^{-2}$ mol/l, S/L = 1/1000).

- Stability of the complexes in solution.

Comparing dependency of U recovery for different systems it was observed that as U complex stability in solution was increasing, pH had less influence on the degree of Uranium extraction (Figure 2); at the same time pH of the solution shifts more during the sorption process.

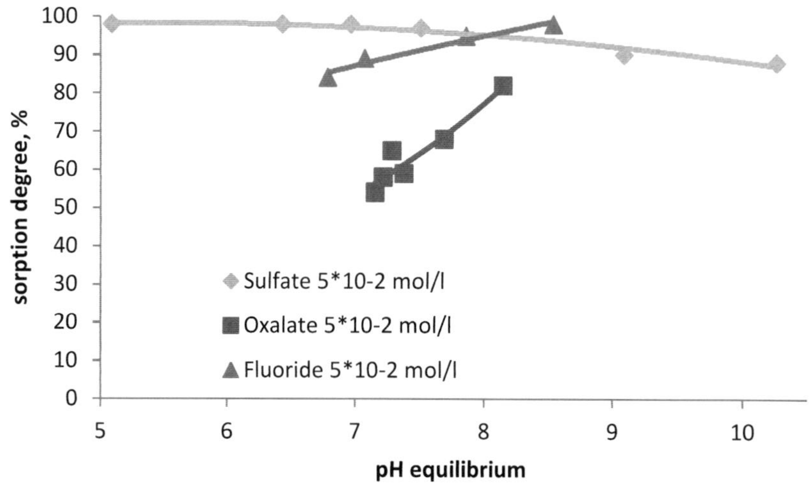

Figure 2. The degree of extraction U and pH dependency. (C(U) = $4{,}2\cdot10^{-5}$ mol/l, time of sorption in the argon atmosphere is 120h, S/L=1/1000, ultrafiltration).

In this study the model of competitive sorption of ions H^+ and positively (or neutral) charged complexes of Uranyl-ion (sulphate, fluoride, oxalate and hydroxo-complexes) on free sorption centers (deprotonated surface hydroxyl group anatase) has been used as a foundation for theoretical description of sorption kinetics. The model based on assumption of the two-step sorption mechanism.

It is assumed that the first step is responsible for the preliminary binding of neutral and positively charged complexes of Uranyl-ion with negatively charged hydroxyl groups of anatase (and binding probability is proportional to the charge of complex ion); the second step provides final binding followed by decomposition of complex ion and binding of UO_2 to the sorption centers.

For the full system description the differential equations of ion equilibrium and material balance were added to the sorption kinetics equations. All ion forms and equilibrium constants have been acquired from well known ion equilibrium in water solutions LLNL database (Lawrence Livermore National Laboratory).

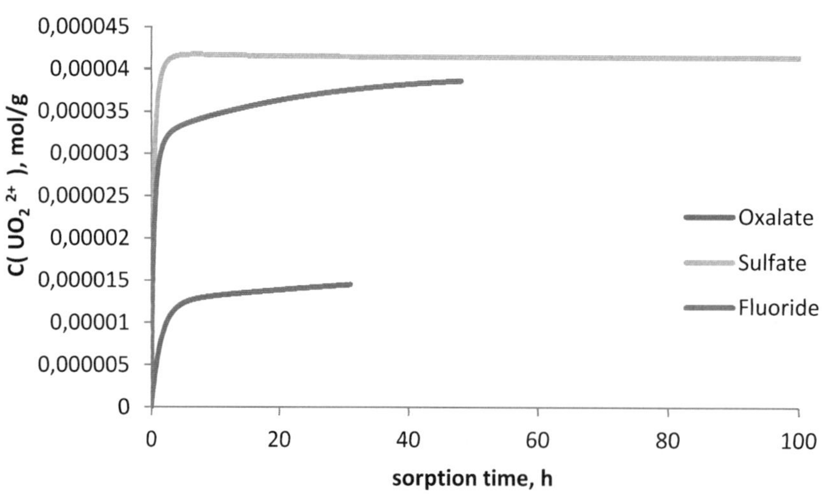

Figure 3. Modeling: change in the concentration of surface UO_2^{2+} groups during sorption of uranium depending on the complexing agent (initial pH 5,4).

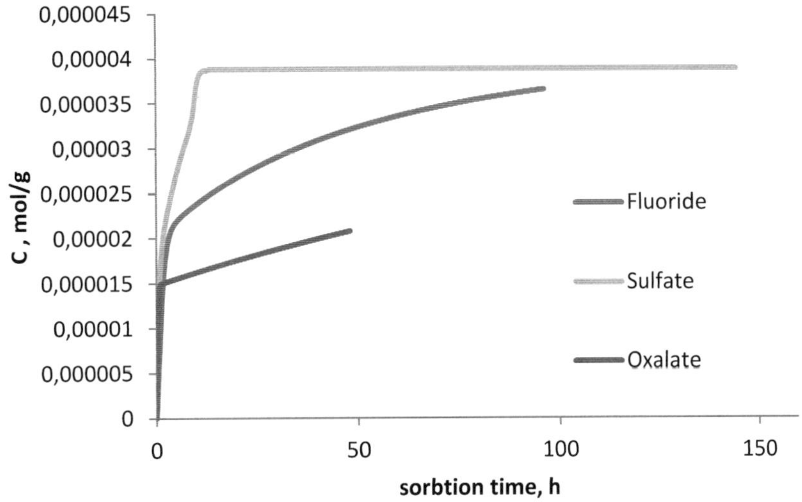

Figure 4. Modeling: change in the concentration of surface UO_2^{2+} groups during sorption of uranium depending on the complexing agent (initial pH 9,9).

The increasing stability of Uranyl-ion complexes in the row oxalates < fluorates < sulfates well known from literature. Modelling of change of concentration of surface $UO_2{}^{2+}$ groups during Uranium sorptions dependant on complexing agent (Figures 3 and 4) clearly shows direct relationship with the complex stability in solution and inverse relationship with the remaining concentration of Uranium in solution (Figure 5). This allows to conclude that proposed model accurately describes influence of complexing agent on the Uranyl-ion sorption.

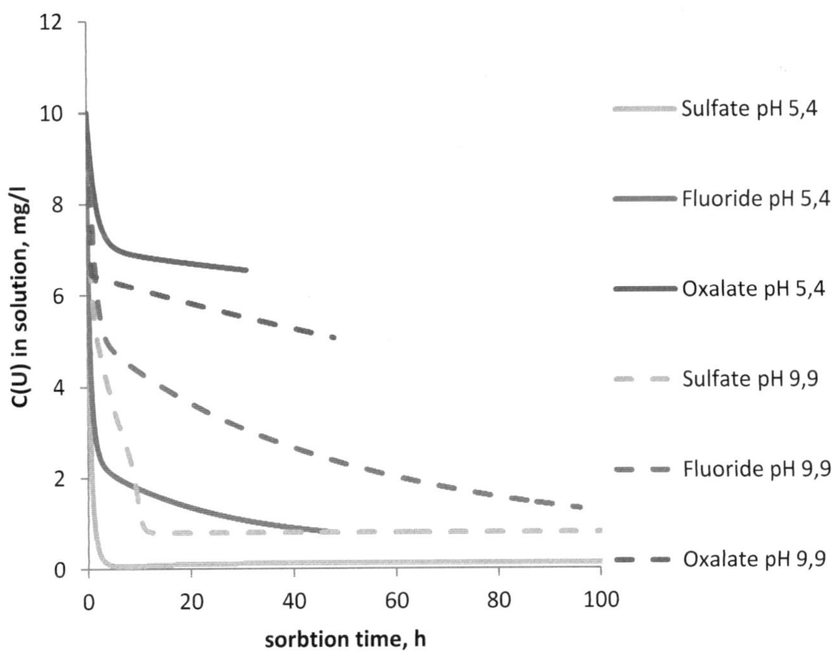

Figure 5. Modeling: Uranium sorption kinetics depending on the complexing agent.

3. Conclusion

Thus we have produced mesoporous nanostructured Titanium Oxides doped by Zirconium ions. The Uranium sorption dependency on the solution pH, type and concentration of complexing agent has been studied. The sorption mechanism were proposed and the sorption process based on them has been modelled.

4. Acknowledgments

The work was supported within the frames of the State Order of the Institute of Chemistry FEBRAS, project № 265-2019-0001.

References

[1] Hygiene standards 2.1.5.2280-07. Additions and changes N 1 to Hygiene standards 2.1.5.1315-03.
[2] ISTC Project Report #KR-072-97; http://vniitf.ru/Karabalta/Result/Pic38_39_1.html
[3] Popa K et al. 2012 *Desalination* **293** 78–86
[4] Perlova O V et al. 2014 *J of Phys. Chem.* **88(6)** 1014
[5] Odoh S O et al. 2012 *Chem. Eur. J.* **18(23)** 7117–7127

Synthetic analogues of natural titanosilicate mesoporous minerals as potential functional materials. Synthesis and application

A I Nikolaev, L G Gerasimova, M V Maslova and E S Shchukina

Federal Research Centre "Kola Science Centre of the Russian Academy of Sciences" Apatity 14 Fersman street.Russia

E-mail: nikol_ai@chemy.kolasc.net.ru

Abstract: Phase formation upon hydrothermal crystallization in the Na_2O-TiO_2-SiO_2-H_2SO_4-H_2O system has been studied in depth. The obtained results provided the basis for the development of technology for new types of ion-exchangers with the structure of the mineral ivanyukite - $Na_4(TiO)_4(SiO_4)_3 \cdot 4\text{-}6\ H_2O$. It was shown that an excessive amount of Na and Si with respect to Ti (mol) and an alkaline medium up to pH 12 affect the rate of the ivanyukite crystallization. The optimal conditions producing the pure ivanyukite phase with designed structure, chemical composition and morphology have been found. The framework composes corner-sharing SiO_4 tetrahedra and TiO_6 octahedra linked through bridging oxygen atoms and provide pore system with channels containing H_2O molecules and exchangeable Na^+cations. The synthetic ivanyukite with wide-pore microporous framework is considered to be perspective cation-exchanger.

1. Introduction

The focus on natural synthetic alkali titanosilicates (Ti-Si) is explained by their versatile properties. Specifically, Ti-Si outperform zeolites, widely used for sorption, not only in terms of higher sorption extent, but also due to thermal stability and selectivity. Determination of conditions for generation of such minerals in nature allows for selection of components and parameters for streamlined synthesis of their analogues with mesoporous particles. Titanosilicates are used for immobilization of radioactive isotopes from aqueous solutions, removal of nonferrous heavy metals from water, separation of gas mixtures and as carrying sorbent agents in medicine. The synthesis of mineral-like Ti-Si is a relative newcomer [1], but in foreign countries many R&D products have been already implemented, e.g. Ti-Si of ETS-4 type [2, 3] or IONSIVIE-911 with a sitinakite mineral texture [4]. New types of Ti-Si minerals – ivanyukites – have been found in the urtites of the Koashva deposit (Kola Peninsula) [5, 6].

2. Main Body

This article describes the ivanyukite technology and its physical and chemical basis. The authors have developed the ivanyukite technology, which in many aspects outperforms the known methods, since it is based on the use of available titanium-containing raw material and innovative processing techniques, providing for high physical and engineering properties of the final products.

2.1. *Research objects*

Titanium salt (ammonium-titanyl-sulfate (ATS) – $(NH_4)_2TiO(SO_4)_2 . H_2O$) obtained from titanite recovered from tails of Khibiny apatite-nepheline ore processing was used for the experiments [7]. Silicone was obtained from soluble glass with mass (%) concentrations: SiO_2 – 31.5; Na_2O – 10. Silicone density: 1.43 g/dm^3. During synthesis the pH value (equal to 11.5) was adjusted by adding NaOH alkali. ATS solution (concentration, mol/L: TiO_2 –1, H_2SO_4 – 1.1–1.5) and alkaline agents have been mixed in set proportions and the mixture was held up for 2 hours. The resulting jellylike precursor was placed into an autoclave and incubated at 200°C within 72 hours. The composition of residual matter resulting from the hydrothermal synthesis was determined with the help of Shimadzu XRD-6000 diffractometer.

Judging from the formula $Na_4(TiO)_4(SiO_4)_3{\cdot}6H_2O$, the molar ratio in ivanyukite Si/Ti and Na/Ti is equal to 0.7 and 1.0 respectively. Synthesis with stoichiometric consumption of components does not allow for generation of ivanyukite. To figure out the optimum synthesis conditions, ensuring generation of ivanyukite, the study was conducted in the system Na_2O-TiO_2-H_2SO_4-SiO_2-H_2O with changes of Na and S molar discharge relative to Ti, which determines the salt mass of the system and induces its "oversaturation" as a key condition for crystal phase formation. Ti, water content and pH value were non-variable.

2.2. *The experimental part*

Based on the study of phase formation in the system, the "system composition – solid phase property" plot was built [8]. The obtained crystallization field consists of nine zones (Figure 1). We noticed that in zone VI limited by the change of Si and Na molar ratios relative to Ti from 3.9 to 5.5 and from 3.5 to 4.8 respectively, the solid phase is generated in the form of two structural versions of ivanyukite: with trigonal crystal structure (T)-$Na_4(TiO)_4(SiO_4)_3{\cdot}6H_2O$ and with cubic structure (C) – $Na_3H(TiO)_4(SiO_4)_3{\cdot}4H_2O$. In zone III, a monophasic sediment crystallizes with the zorite structure – $Na_6Ti_5Si_{12}O_{34}(O,OH)_5{\cdot}11H_2O$. In other zones of the system, generation of polycomponent sediments containing various amounts of titanosilicatephases (natisite, sitinakite, zorite, ivanyukite, as well as TiO_2anatase and quartz) takes place.

XRD patterns of samples in zone VI obtained under conditions specified in Table 1 are almost identical (Figure 2). Structural varieties of ivanyukite include slight displacement of the main peak (100) for the samples with mixed structure 1 and 2. The same samples are noted for a much higher peak intensity (211) as compared to sample 3, which is an evidence of its monophasic composition. Surface behavior was determined by nitrogen adsorption/desorption (BET method) using TriStar 3020 analyzer (Table 1).

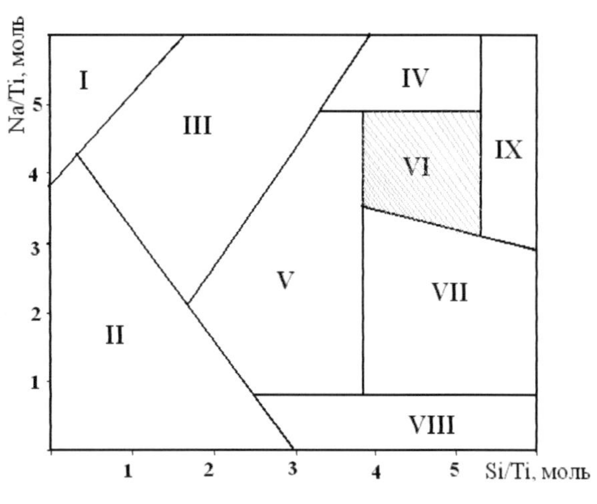

Figure 1. Crystallization field of alkali titanosilicate solids in Na_2O-TiO_2-H_2SO_4-SiO_2-H_2O system

High BET surface area of ivanyukite (135–150 m^2/g), its porosity (gross pore volume: 0.75–0.81 cm^3/g) and availability of fluent off-scaffold cations Na^+ and water molecules positively affect the speed and efficiency of exchange processes taking place in sorption systems. Sorption capacity (E) of ivanyukite was determined for sample 2 relative to cations Cs^+, Sr^{2+} and Co^{2+}, mg eq/g: E_{Cs}– 2.9; E_{Sr} – 4.5; E_{Co} – 3.2. A patent for the invention was received [9].

Table 1. Surface behavior of ivanyukite samples

Sample No.	Consumption of agents during ivanyukite synthesis in zone VI	Surface behavior		
		SG, m^2/g	Vpore, cm^3/g	Dpore, nM
1	4.1 Na:Ti:5.0Si:160H_2O ivanyukite-*T*+ ivanyukite-C	150.2±1.3	0.75	15.0
2	4.5Na:Ti:4.2 Si:160H_2O ivanyukite-*T*+ ivanyukite-C	137.6±1.0	0.78	16.0
3	3.9Na:Ti:5.5Si:160H_2O ivanyukite-C	134.9±1.1	0.81	15.5

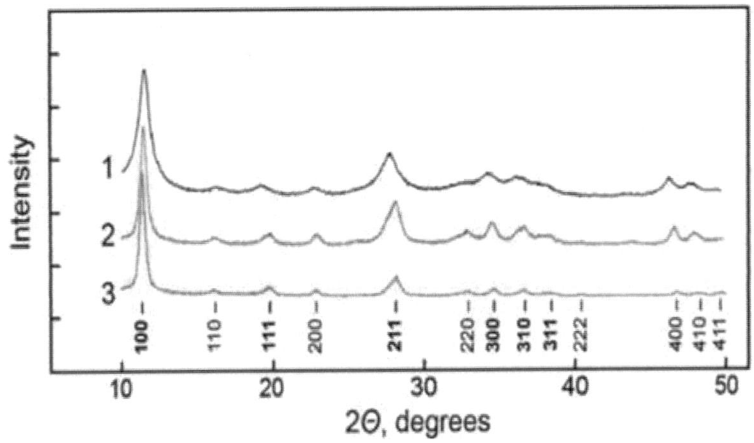

Figure 2.XRD patterns of samples

Titanosilicate precursor granulation technique using no additional agents and providing for high stability of granules during long-term dynamic impact, while maintaining their sorption characteristics, was developed (Figure 3).

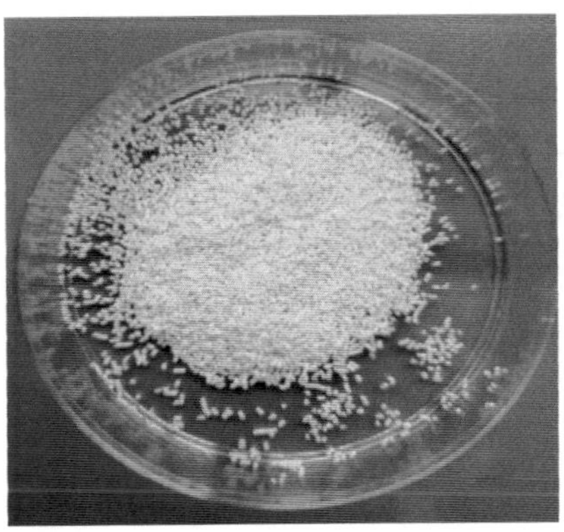

Figure 3. Granular ivanyukite. Grain size: 1–1.3 mm

3. Conclusion

Derivatization methods of other structural types of titanosilicates using sulphate titanium salts are currently being developed. In particular, a diagram of stepwise synthesis of layered titanosilicate AM-4 ($Na_3NaTi_2[Si_4O_{14}]\cdot 2H_2O$) with a structure similar to lintisite and tundrite minerals was proposed. It has been established that in acidic media AM-4 is converted into a layered titanosilicate SL3 which selectively extracts monovalent cations allowing for its use for silver recovery from complex process solutions and ^{137}Cs from LRW [10]. Argentic form SL3-Ag exhibits photocatalytic properties.

Spent radioactive sorbent agent can be turned into a stable titanate ceramic corresponding to a solid highly reactive porcelaneous mass, similar to Synroc ceramics by composition [11].

The reported study was financially supported by RFBR according to the research project 18-29-121039 and was partly supported by the theme of state assignment of AAAA-A17-117020110035-5

References

[1] Wang X and Jacobson A J. 1999 Crystal structure of the microporoustitanosilicate ETS-10 refined from single crystal X-ray diffraction date *Chem. Commun* **11** 973 –4.

[2] Noh S H, Kim S D and Chung Y J 2006 The effects of (Na + K)/Na molar ratio and kinetic studies on the rapid crystallization of a large pored titanium silicate, ETS-10 using cost efficient titanium oxysulfate, $TiOSO_4$ under stirring *Micropor. Mezpor. Mat.* **88** (1-3) 197-204.

[3] Ji Z, Yilmaz B, Warzywoda J and Ji A S 2005 Hydrothermal synthesis of titanosilicate ETS-10 using $Ti(SO_4)_2$ *Micropor. Mesopor. Mat.* **81** (1-3) 1 – 10.

[4] Perovsky I A and Burtsev I N 2013 Hydrothermal synthesis of sitinakite on the basis of leucoxene of the Yarega Field *Bulletin of the Komi Research Center of the Russian Academy of Sciences. Syktyvkar* **3**. 16-9.

[5] Yakovenchuk V N, Nikolaev A P, Selivanova E A, Pakhomovsky Y A, Korchak J A, Spiridonova D V, Zalkind O A and Krivovichev S V 2009 Ivanyukite-Na-T, ivanyukite-Na-C, ivanyukite-K, and ivanyukite-Cu: New microporoustitanosilicates from the Khibiny massif (Kola Peninsula, Russia) and crystal structure of ivanyukite-Na-T *American Mineralogist* **94** 1450–8.

[6] Yakovenchuk V N, Krivovichev S V and Pakhomovsky Ya A 2012 Microporous titanosilicates of the lintisite-kukisvumite group and their transformation in acidic solutions *Minerals as Advanced Materials IIed SV Krivovichev Springer-Verlag, Berlin-Heidelberg* 229–238.

[7] Gerasimova L G, Nikolaev A I, Maslova M V and Shchukina E S 2017 Synthesis of a titanium (IV)-based sorbent and potentialities of its usage for extracting cations of non-ferrous metals *Non-ferrous Metals* **2** 32-36.

[8] Pavel C C, Vuono D, Catanzaro L, De Luca P, Bilba N, Nastro A and Nagy J B 2002 Synthesis and characterization of the microporous titanosilicates ETS-4 and ETS-10 *Micropor. Mesopor. Mat.* **56** (2) 227-39.

[9] Gerasimova L G, Nikolaev A I and Shchukina E S The method of obtaining sodium-containing titanosilicate. Patent RF № 2568699, 2015.

[10] Kalashnikova G O Obtaining of new sorbents of cesium, silver and iodine by the reversible transformation of lintisite-like titanosilicates (synthesis, properties and prospects for use) *Extended abstract of PhD dissertation (Engineering)*, Apatity. 2016. 23 p.

[11] Britvin S N, Gerasimova L G, Ivanyuk G Yu, Kalashnikova G O, Krzhizhanovskaya M G, Krivovivhev S V, Mararits V F, Nikolaev A I, Oginova O A, Panteleev V N, Khandobin V A, Yakovenchuk V N and Yanicheva N Yu, 2016 Application of titanium-containing sorbents for treating liquid radioactive waste with the subsequent conservation of radionuclides in Synroc-titanate ceramics. *Theoretical Foundations of Chemical Engineering* **50** (4) 599–606.

Synthesis and spectral-optical properties of composite materials based on high-silica porous glasses doped by iodides of silver and erbium

M A Girsova[1], T V Antropova[1], G F Golovina[1], I N Anfimova[1], L N Kurilenko[1], M Yu Arsent'ev[2]
[1]Laboratory of Physical Chemistry of Glass, Grebenshchikov Institute of Silicate Chemistry RAS, Saint-Petersburg 199034, Russia
[2]Laboratory of Research of Nanostructures, Grebenshchikov Institute of Silicate Chemistry RAS, Saint-Petersburg 199034, Russia

E-mail: girsovamarina@rambler.ru

Abstract. Composite materials (CMs) doped by iodides of silver and erbium were synthesized on the base of the thermally modified high-silica porous glasses (PGs) with small additions of P_2O_5 and fluoride ions. CMs were manufactured by two-stage impregnation of the PGs first with $AgNO_3$ aqueous solution (with or without $Er(NO_3)_3$), next in the halide salt (0.6 M KI) aqueous solution. Then the samples of composites were heat treated at the temperatures from 120 to 870 °C up to their sintering and closing of the pores. The structure of composites was studied by X-ray powder diffraction (XRD) technique. According to XRD data the CMs doped by iodides of silver and erbium contain the AgI phase. Spectral-optical properties of CMs were investigated by UV-VIS-NIR spectroscopy. Optical density spectra in UV-VIS region have the bands corresponded with the absorption of silver molecular clusters. Peaks at 377, 448, 450–453, 485, 611–622, 647, 668–684 nm answered to the absorption of erbium ions were found. According to NIR transmission spectra, bands caused by the absorption of Si–OH, B–OH and Na–OH groups are found.

1. Introduction

It is known that the simultaneous presence of silver nanoparticles and rare earth (RE) ions in the material leads to energy transfer processes, where silver and RE ions can be sensitizers [1, 2]. Besides that the role of RE ions depends on their concentration in material [3].

2. Experimental

In this work, composite materials (CMs) were synthesized based on matrices from thermally modified porous glasses (PGs) with small additions of P_2O_5 and fluoride ions (PG-NFF) [4, 5]. Thermal modification of PG-NFF was carried out by its heating within wide temperature (T) range from 120 °C (PG-NFF-120) to 700 ° C (PG-NFF-700) in accordance with the procedure first proposed in [4]. It was found that with PG's temperature increasing, the pore specific surface area (SA) decreases from ~ 65 to ~ 37 m^2/g, the average pore diameter D increases from ~ 10 to ~ 17 nm while maintaining the porosity W ~ 0.30 cm^3/cm^3. The values of W were determined by gravimetric method, and SA by thermal desorption of nitrogen at 77 K (BET) method. The values of D were calculated in accordance with the procedure described in [4].

Additional heat treatment of PG-NFF-120 samples at T = 870 ± 5 ° C leads to complete pore collapse and the formation of quartz-like glass QG-NFF. According to chemical analysis, the basic porous matrices PG-NFF-120 and QG samples contain (wt. %) (0.46–0.55) Na_2O, (6.42–6.85) B_2O_3, (92.56–93.09) SiO_2, (0.03–0.04) |F|, ≤ 0.01 P_2O_5 [5].

Synthesis of the CMs was carried out by two-stage impregnation of PG-NFF. At first, PG-NFF matrices (in the form of the rectangular plane-parallel plates in size (5–15) x (5–25) x (1.5±0.15) mm^3) were placed in 100 mg/ml $AgNO_3$ aqueous solution (with or without the presence of sensitizers, such as 10 mg/ml $Er(NO_3)_3$). Next the samples were immersed in a halide salt (0.6 M KI) aqueous solution. Then CMs were heat treated at the temperatures from 120 to 870°C for 15–160 minutes. Such heat treatment of CMs and PGs leads to their sintering with closing of the pores, i.e. a monolithic high silica material is formed.

In the CM samples of series (50 Ag, 100 Ag, 100Ag/10Er) based on PG-NFF-120 matrices doped with silver iodide with or without erbium iodide, the content of silver and alkali metals (wt. %) is (0.45–0.56) Na_2O, (0.49–0.57) K_2O, (0.68–1.02) Ag_2O as determined by flame photometry on an iCE of 3000 Series spectrophotometer.

XRD study of the CMs (on example 100Ag/10Er series based on PG-NFF-120 matrix, heat-treated in air at 120 °C) was carried out on D8-Advance "Bruker" X-ray diffractometer (monochromatic CuKα radiation, operating voltage of 40 kV at a current of 40 mA in the angle range 2θ = 20–60 °) using the database ICDD-2006 on the samples previously powdered.

Synthesized CMs were investigated by optical spectroscopy methods in UV-VIS-NIR spectral regions, depending on composition and temperature of heat treatment of them (120–870 °C), as well as of the PG-NFF matrices. Samples of PG-NFF with different thermal history and QG-NFF were chosen as the objects of comparison. The measurements were carried out at room temperature on the samples with a thickness of 1.00 and 1.50 mm (± 0.15 mm). The optical density spectra were obtained in the spectral range 270–1100 nm on SF-2000 spectrophotometer, and FTIR transmission spectra in the range 7500–4000 cm^{-1} on FSM-2211 spectrophotometer.

3. Results and discussion

According to XRD data, crystalline phase of the hexagonal modification AgI (No. 01-078-1614) with average crystallite size of 26 nm was estimated by the Scherer method (Fig. 1).

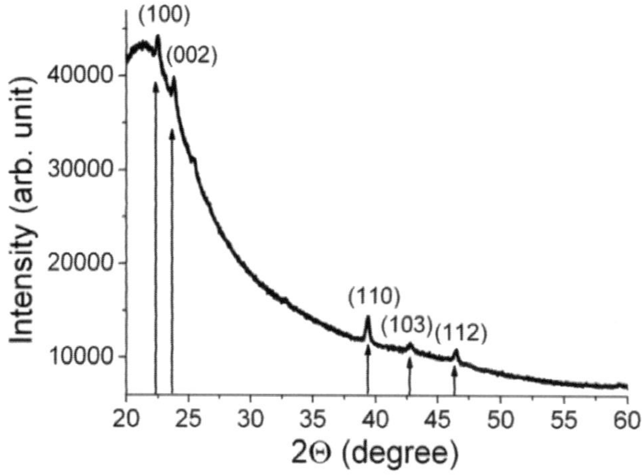

Fig.1. Diffractogram of the CM sample of series 100Ag/10Er based on the PG-NFF-120 matrix

Fig. 2 shows the optical density spectra of thermally modified porous glasses in the spectral region 270–1100 nm. The presence of UV absorption which is possibly associated with silicon defect centers [6] in PG-NFF matrices, has been established.

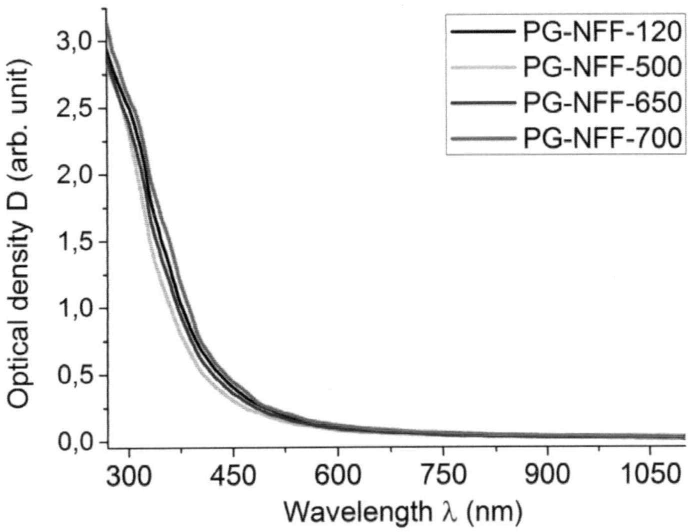

Fig. 2. Optical density spectra of thermally modified PGs

Figure 3 and Figure 4 show the optical density spectra of the CMs of series 50Ag, 100Ag and 100Ag/10Er samples obtained on the PG-NFF-120 matrices, depending on their heat treatment temperature.

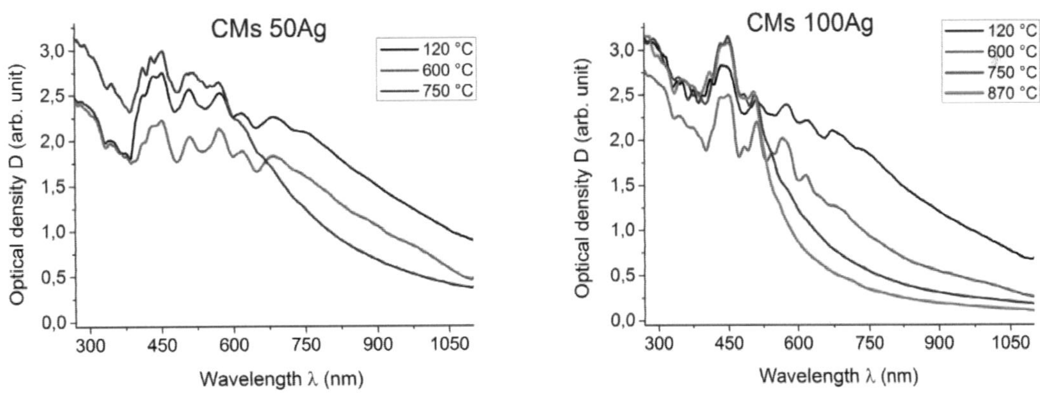

Fig. 3. Optical density spectra of the CMs of series 50Ag and 100Ag depending on the temperature of their heat treatment (120–870 °C)

The spectral dependences revealed absorption bands at the wavelength 340–346 nm (~ 3.58–3.65 eV), 351–354 nm (~ 3.50–3.53 eV), 372–385 nm (~ 3.22–3.33 eV), 483–485 nm (~ 2.56–2.57 eV), 504–512 nm (~ 2.42–2.46 eV), 544 nm (2.28 eV), 611–622 nm (~ 1.99–2.03 eV), which are corresponded with the absorption of silver molecular clusters [7, 8]. It has been established that in most of the CM samples the peaks appear at the wavelengths 395–455 and 564–575 nm, which may

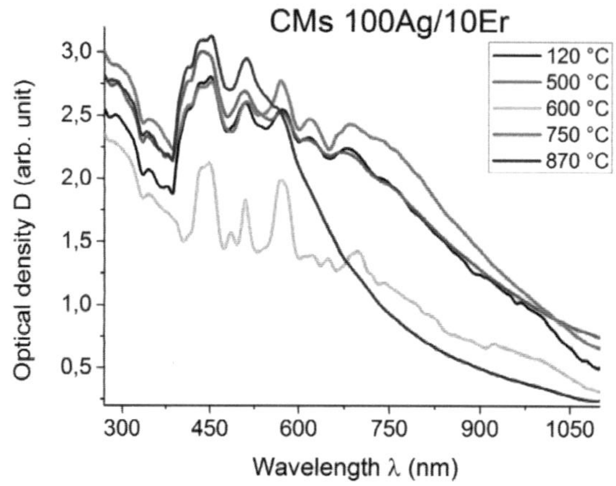

Fig. 4. Optical density spectra of the CM 100Ag/10Er depending on the temperature of their heat treatment (120–870 °C)

indicate the formation of AgI nanocrystals and interspersed silver nanoparticles with plasmon resonance [9–12]. With increase in the temperature of heat treatment of CMs from 120 to 870 ° C, a change in the position of the absorption bands (or their disappearance) associated with silver presence is observed at the spectra. The peaks at the optical density spectra of CMs at 377, 448, 450–453, 485 and 611–622, 647, 668–684 nm can also be associated with the absorption of erbium ions that relate to transitions $^4I_{15/2}\rightarrow{}^4G_{11/2}$, $^4I_{15/2}\rightarrow{}^4F_{3/2}$, $^4I_{15/2}\rightarrow{}^4F_{5/2}$, $^4I_{15/2}\rightarrow{}^4F_{7/2}$ and $^4I_{15/2}\rightarrow{}^4F_{9/2}$ respectively [12–16].

Figure 5 and Figure 6 show typical transmission spectra of the CMs of series 50 Ag and 100Ag/10Er based on PG-NFF-120 matrices in the near infrared region. Most of the CM samples exhibit absorption bands at the frequency range 7332–7315 cm^{-1} (stretching vibrations of OH and Si–OH groups), 7189–7116 and 5281–5261 cm^{-1} (absorption of OH groups and water molecules), at 4686–4646 cm^{-1} (stretching vibrations of B–OH groups), 4536–4503 cm^{-1} (deformation vibrations of Si–OH groups, vibrations of OH and Na–OH groups), 4453–4437 and 4192–4155 cm^{-1} (vibrations of OH and Si–OH groups) (see the overview in [5]). It should be noted that absorption bands in the

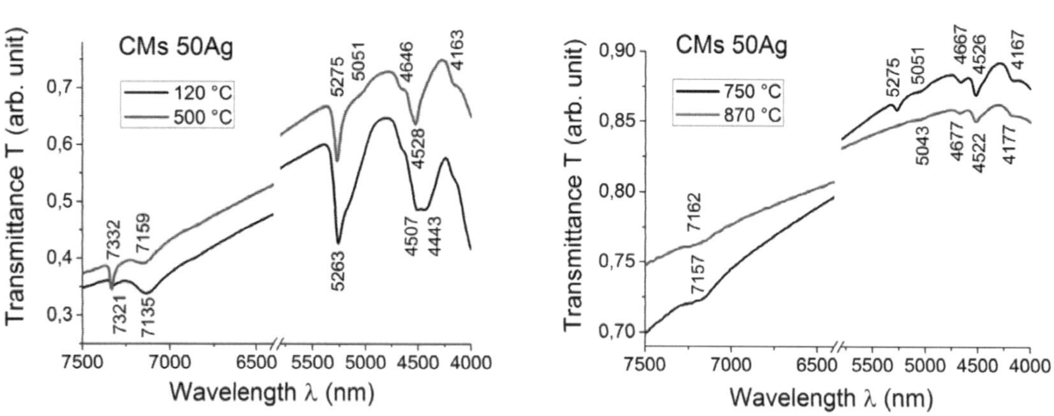

Fig. 5. Transmission spectra of the CM 50 Ag depending on the temperature of their heat treatment

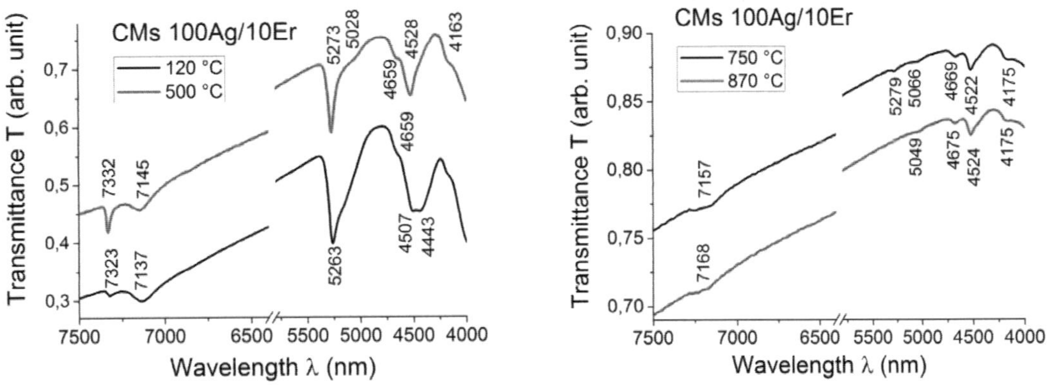

Fig. 6. Transmission spectra of the CM 100Ag/10Er depending on the temperature of their heat treatment

frequency range 6583–6486 cm^{-1} (~ 1519–1542 nm), characteristic for erbium ions [12–15], were not detected.

4. Conclusion

XRD spectra of the synthesized CM's powders show that they contain AgI phase of the hexagonal modification. According to NIR transmission spectra, CMs contain Si–OH, B–OH, Na–OH groups. Optical absorption spectra of the CM 100Ag/10Er show the bands at 377, 448, 450–453, 485 and 611–622, 647, 668–684 nm corresponded with the absorption of erbium ions (transitions $^4I_{15/2}\rightarrow{}^4G_{11/2}$, $^4I_{15/2}\rightarrow{}^4F_{3/2}$, $^4I_{15/2}\rightarrow{}^4F_{5/2}$, $^4I_{15/2}\rightarrow{}^4F_{7/2}$ and $^4I_{15/2}\rightarrow{}^4F_{9/2}$ respectively).

Acknowledgments

This work was carried out within the scope of the State Assignment of IChS RAS for the Program of Fundamental Scientific Research of State Academies of Sciences for 2013-2020 (State registration number AAAA-A19-119022290087-1).

The authors thank L. F. Dikaya, L. A. Doronina, T. G. Kostyreva, and E. A. Semenova for chemical analysis of the glasses.

References

[1] Eichelbaum M and Rademann K 2009 *Adv. Funct. Mater* **19** 2045,
[2] Strohhöfer C and Polman A 2002 *Applied Physics Letters* **81** 1414,
[3] Auzel F 1990 *Journal of Luminescence* **45** 341,
[4] Antropova T V, Anfimova I N and Golovina G F 2009 *Glass Physics and Chemistry* **35** 572,
[5] Girsova M A and Golovina G F 2018 *Glass Physics and Chemistry* **44** 569,
[6] Sokolov V O and Sulimov V B 1994 *Phys. Stat. Sol. B* **186** 185,
[7] Lecoultre S, Rydlo A, Buttet J, Félix C, Gilb S and Harbich W 2011 *J. Chem. Phys.* **134** 184504
[8] Sgibnev Y M, Nikonorov N V and Ignatiev A I 2016 *Journal of Luminescence* **176** 292,
[9] Wackerow S, Seifert G and Abdolvand A 2011 *Optical Materials Express* **1** 1224,
[10] Andreeva O V, Sidorov A I, Stasel'ko D I and Khrushcheva T A 2012 *Physics of the Solid State* **54** 1293,
[11] Obraztsov P A, Nashchekin A V, Panfilova A V, Brunkov P N, Nikonorov N V and Sidorov A I 2013 *Physics of the Solid State* **55** 1272,

[12] Moustafa S Y, Sahar M R and Ghoshal S K 2017 *Journal of Alloys and Compounds* **712** 781,
[13] Carnall W T, Fields P R and Pajnak K 1968 *J. Chem. Phys.* **49** 4424,
[14] Kaky K M, Lakshminarayana G, Baki S O, Lira A, Caldiño U, Meza-Rocha A N, Falcony C, Kityk I V, Taufiq-Yap Y H, Halimah M K and Mahdi M A 2017 *Optical Materials* **69** 401,
[15] Kostka P, Kabalci I, Tay T, Gladkov P and Zavadil 2017 *Journal of Luminescence* **192** 1104,
[16] Judd B R 1962 *Phys. Rev.* **127** 750.

Nonlinear viscoelastic properties of nanosuspensions

T S Dembelova, B B Badmaev, D N Makarova and Ye D Vershinina

Institute of Physical Materials Science, Siberian Branch of the Russian Academy of Sciences, 670047, Ulan-Ude, Russia

E-mail: tu_dembel@mail.ru

Abstract. This paper studies the dependence of the shear elasticity modulus and the coefficient of dynamic viscosity of the suspensions on the magnitude of the shear strain of nanoparticle suspensions in a polymer liquid. As the dispersion medium, polyethylsiloxane liquid PES-2 was used, the solid dispersed phase - silica nanoparticles and fullerenes. We used the acoustic resonance method for determining the complex shear modulus of liquids using a piezoquartz resonator. Studies have shown nonlinear behavior of viscoelastic parameters of suspension samples depending on shear stress.

1. Introduction

There are many methods for studying the viscoelastic properties of liquids and multiphase media. Resonance methods for measuring the complex shear modulus of liquids are based on measuring the resonance characteristics of an oscillatory system. The elastic forces arising from the shear deformation of the liquid increase the resonant frequency of the vibrator, and the viscous ones increase the attenuation of the oscillatory system. It is well known that linear viscoelasticity is observed only at small deformations. With increasing strain, the resonance curve of the oscillatory system deforms. This paper presents a study of viscoelastic properties of colloidal suspensions of silica dioxide nanoparticles and fullerenes using a resonance method in dependence on the amplitude of shear vibrations. Polyethylsiloxane liquid was used as a base fluid. The wide range of working temperatures (from -60 to +150 °C) allows to widely using them in modern technology. Silica dioxide nanopowder SiO_2 obtained by gas-phase synthesis [1], and fullerenes C_{60}, purchased from Company "NeoTechProduct", were used for suspensions.

2. Method

The viscoelastic properties of nanoparticle suspensions are studied by the acoustic resonance method using a piezoelectric crystal [2-4]. A piezoelectric crystal in the shape of a rectangular bar performs compression and tension oscillations at the fundamental resonant frequency. The liquid is applied to one end of the horizontal surface of the piezoelectric quartz and covered with a solid fused silica cover-plate. A certain crystal cut (X-18.5 °), having a zero Poisson's ratio on the working face, allows one to obtain purely shear deformations in the studied liquid layer. In this case, standing shear waves are established in it. Depending on the thickness of the liquid layer, the parameters of the resonance curve of the system change. Under the assumption that the cover-plate is loosely bonded to the piezoquartz through the liquid layer and, therefore, is at rest and the thickness of the interlayer is much less than the shear wavelength in the liquid, the complex shear modulus G^* and the tangent of the mechanical loss angle $\tan\theta$ of the investigated liquid will be determined by the following expressions:

$$G^* = \frac{4\pi^2 M f_o \Delta f^* H}{S}, \qquad \tan\theta = \frac{G''}{G'} = \frac{\Delta f''}{\Delta f'}, \tag{1}$$

where $G^* = G' + iG''$ is the complex shear modulus of the fluid; $\Delta f^* = \Delta f' + i\Delta f''$ is the complex shift of the resonant frequency; M is the mass of the piezoelectric quartz; S is the area of the cover-plate base; f_o is the resonant frequency of the piezoelectric crystal; H is the thickness of the liquid layer. The resonant frequency of the piezoelectric crystal was 73.2 kHz, the mass was 6.82 g, and the area of the base of the cover-plate was 0.2 cm^2.

The imaginary shift of the resonant frequency is equal to the change in the attenuation of the oscillatory system: $\Delta f'' = \frac{\Delta\alpha}{2}$, where $\Delta\alpha$ is the change in the width of the resonance curve of the piezoelectric quartz. Knowing the change in the resonance frequency and the width of the resonance curve, it is possible to determine the complex modulus of shear of the liquid.

3. Results and discussion

According to theory of a nonlinear oscillatory system it is known that its properties are determined through the parameters of the resonance curves. In the case of nonlinearity of the shear elasticity of liquids, the resonance curve of the oscillatory system of the piezoelectric quartz — the interlayer of the liquid — the cover-plate will deform with an increase in the amplitude of the oscillation of the piezoelectric quartz becoming more asymmetric, and the properties of the oscillatory system will be determined through the parameters of the resonance curves. The vibration amplitude of a piezoelectric crystal can be determined using a method based on the principle of the Fabry-Perot interferometer, where the optically polished end face of a piezoelectric crystal is used as one of the mirrors [5]. The obtained dependence of the amplitude of the oscillation of piezoelectric crystal A on the voltage removed from it makes it possible to measure the amplitude with sufficiently high accuracy. The A/H ratio can serve as a measure of angular deformation.

Figure 1 shows the change in the shape of the resonance curves with increasing values of the driving force oscillating the piezoelectric quartz for the hexadecane layer studied in [6] by the acoustic resonance method. Curve 1, taken at low vibration amplitudes, is symmetric and corresponds to the region of linear elasticity. As the driving force increases, the resonance curve becomes steeper from the low frequency side and further exhibits instability. The middle line of the resonance curve 3 actually depicts the dependence of shear elasticity on the angle of deformation.

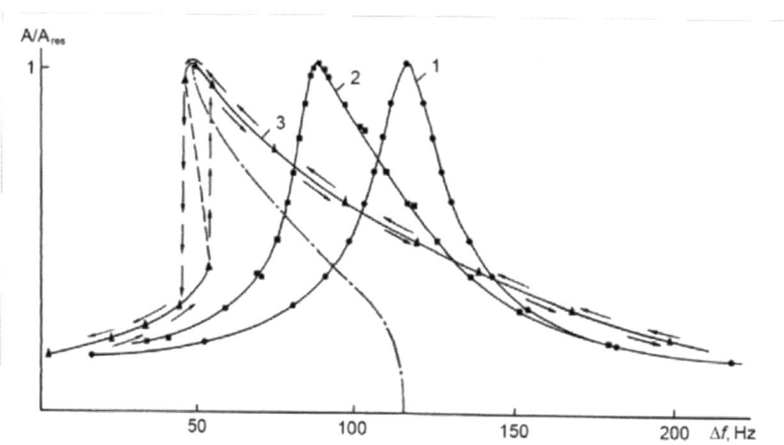

Figure 1. Resonance curves for various driving forces with a maximum amplitude of oscillation: 1 – 10 Å; 2 – 100 Å; 3 – 300 Å [6].

IOP Conf. Series: Materials Science and Engineering **704** (2019) 012005 doi:10.1088/1757-899X/704/1/012005

Figure 2 shows the dependences of the real and imaginary components of the shear modulus calculated by formula (1) on the shear angle for samples of silica suspensions with a nanoparticle size of 100 nm with a concentration of 1.25 wt.% and fullerenes with a concentration of 1.5 wt.% in a polymer liquid PES-2. Suspensions are obtained by ultrasonic method. For ease of analysis, the results are presented depending on the value $(A/H)^{1/2}$. At small deformation angles, the components of the complex shear modulus are constant, then, with an increase in the shear angle, G′ decreases, and G″ passes through a maximum. Similar curves for pure liquids were obtained in [5,7]. It can be assumed that the liquid has an equilibrium supramolecular structure and, at small angles of shear deformation, a region of linear elasticity is detected. At a certain critical shear stress P_c, the equilibrium structure begins to collapse and its viscoelastic properties change. The critical shear stress corresponds to a certain critical shear angle φ_c, which can be determined from the data in Fig. 2.

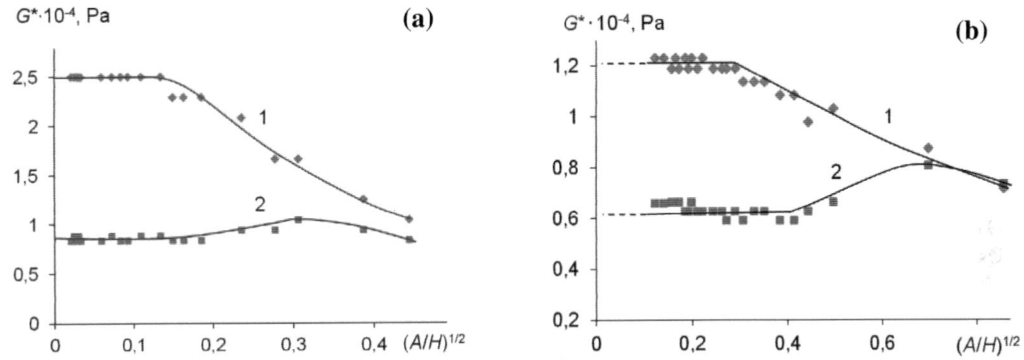

Figure 2. Experimental dependences of real (1) and imaginary (2) parts of shear modulus on the shear deformation angle for suspensions: **(a)** SiO_2/PES-2, 1.25 wt.%; **(b)** C_{60}/PES-2, 1.5 wt.%.

In the region of linear elasticity, G′ and tanθ for the suspensions studied are G′ = $2.5\cdot10^4$ Pa and tanθ = 0.35 and G′ = $1.2\cdot10^4$ Pa, tanθ = 0.55, respectively.

Figure 3 shows the graphs of effective viscosity versus shear strain angle for these suspensions. Viscosities are calculated using Maxwell's rheological model by the formula:

$$\eta_{eff} = \frac{G'(1+\tan^2\theta)}{2\pi f_o \tan\theta}. \tag{2}$$

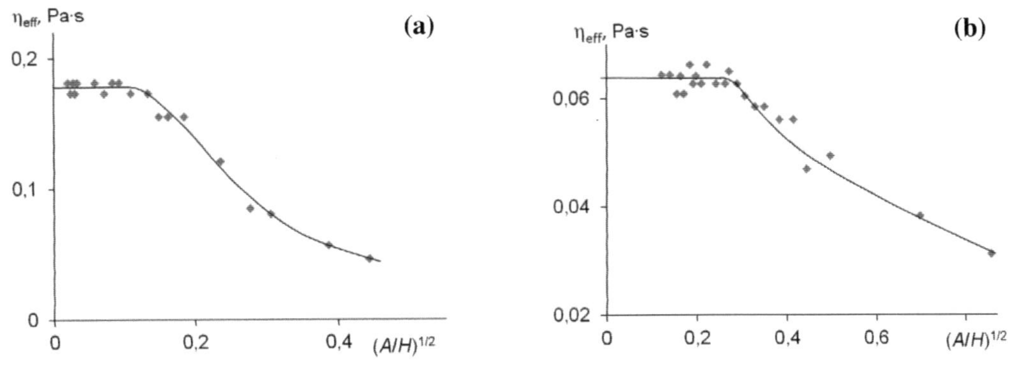

Figure 3. Dependence of effective viscosity on the shear deformation angle for suspensions: **(a)** SiO_2/PES-2, 1.25 wt.%; **(b)** C_{60}/PES-2, 1.5 wt.%.

The suspensions are characterized by a constant increased viscosity when shear angles smaller than the critical value φ_c. At these region effective viscosities of suspensions are higher than the value of viscosity for base liquid, which have been determined by a rotation method using a rheometer Physica Anton Paar MCR302. It equals to 0.01 Pa·s. With a further increase in the shear angle, the structure is destroyed and the viscosity decreases to the smallest value.

Figure 4 shows the dependence of the shear stress in the suspension layer on the magnitude of the shear strain, the dashed line shows the linear dependence of the shear stress. Critical shear stress when the linearity of the dependence is violated, for the suspension SiO_2/PES-2 is $P_c = 0.45 \cdot 10^3$ Pa and for the C_{60}/PES-2 is $P_c = 1.04 \cdot 10^3$ Pa. The nonlinearity of the shear elasticity of the samples with an increase in the angle of deformation is clearly visible in the figure; moreover, for a silicon dioxide suspension having a lower value of the tangent of the angle of mechanical losses, the nonlinearity is more pronounced.

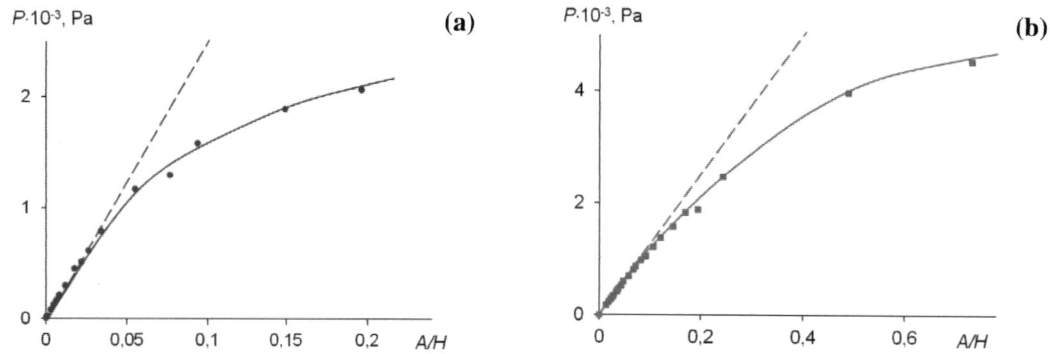

Figure 4. Dependence of shear stress on the shear strain for suspensions: **(a)** SiO_2/PES-2, 1.25 wt.%; **(b)** C_{60}/PES-2, 1.5 wt.%. The dashed line corresponds to a linear dependence.

4. Conclusion

Thus, the shift of the resonance frequency of the vibrational system of piezoelectric quartz — the suspension layer — the overlay depends on the amplitude of the shear strain, which indicates the nonlinearity of the viscoelastic properties of the suspensions under study.

Acknowledgment
This work was partially supported by the RFBR grant No. 18-48-030020 r_a.

References

[1] Patent 2067077 Russian Federation, IPC C01B33/18
[2] Badmaev B B, Bazaron U B, Derjaguin B V and Budaev O R 1983 *Physica B+C* **122** 241–5
[3] Badmaev B B, Dembelova T S and Damdinov B B 2013 *Viscoelastic properties of polymer liquids* (Ulan-Ude: Publisher of the Buryat Scientific Center, SB RAS)
[4] Badmaev B, Dembelova T, Damdinov B, Makarova D and Budaev O 2011 *Colloids Surf.* A **383** 90–4
[5] Badmaev B B, Dembelova T S, Makarova D N and Gulgenov C Z 2017 *Tech. Phys.* **62** 14–7
[6] Bazaron U B, Deryaguin B V, Zandanova K T and Lamazhapova Kh D 1981 *Zhurnal fizicheskoi khimii* **55** 2812-6
[7] Badmaev B B, Makarova D N, Damdinov B B, Sanditov D S and Dembelova T S 2014 *Russ. Phys. J.* **57** 742-8

IOP Conf. Series: Materials Science and Engineering **704** (2019) 012006 doi:10.1088/1757-899X/704/1/012006

The temperature coefficient delay time of the Rayleigh waves

I G Simakov, Ch Zh Gulgenov and S B Bazarova*

Institute of Physical Materials Science, Siberian Branch of the Russian Academy of Sciences, 670047, Ulan-Ude, Russia

*E-mail: baz_say@mail.ru

Abstract. The effect of variation of temperature of adsorbing surface and humidity of gas medium on velocity of surface acoustic waves in layered system "adsorbed water - lithium niobate" has been investigated. The dependence of temperature coefficient delay time of the Rayleigh waves in system "adsorbed water - lithium niobate" on degree of moisture content of gas medium was considered.

Acoustoelectronic devices hade practical application in modern radio signal processing systems. The main element of acoustoelectronic devices is the SAW delay line. The SAW delay line is a piezoelectric substrate on which surface acoustic waves of Rayleigh type (SAW) propagate. To generate and register of SAWs on a polished surface of a piezoelectric substrate – acoustic line, two (or more) interdigital transducers are formed. Acoustoelectronic devices often work under destabilizing effects ambient gas.

In a humid gas environment, a thin layer of water forms as a result of adsorption on the working surface of the acoustic line of an acoustoelectronic device. The liquid layer on the surface of the acoustic line result in changes in the conditions for the propagation of SAW and consequently, to a change in the signal parameters at the output of the acoustoelectronic device.

The thickness of the adsorption layer depends on the temperature and degree of humidity of the vapor-gas medium. Influence of temperature and humidity of gas medium leads to change of conditions of surface waves propagation. This results in a change in attenuation and surface wave velocity. In addition, the linear dimensions of the acoustic line change. This work is devoted to the study of the complex effect of temperature and humidity of the environment on the temperature coefficient delay time of Rayleigh waves.

The SAW velocity change and linear dimensions of the acoustic line leads to a change in the temperature coefficient delay time (TCD) $\zeta = \tau^{-1}\partial\tau/\partial T$, which in the absence of an adsorption layer of water has the following form:

$$\zeta = \frac{1}{\tau}\frac{\partial \tau}{\partial T} = \frac{1}{L}\frac{\partial L}{\partial T} - \frac{1}{V}\frac{\partial V}{\partial T} = \alpha - \frac{1}{V}\frac{\partial V}{\partial T}$$

where τ – delay time of surface acoustic waves, L – length of the acoustic line, V – velocity of surface acoustic waves. TCD includes the temperature coefficient surface wave velocity change $\partial V/(V\,\partial T)$ and the coefficients of linear thermal expansion α [2].

Change of delay time of acoustic signal in case of variation of substrate (acoustic line) temperature must be investigated for investigation of TCD dependence on vapor-gas medium degree of humidity.

As a result of adsorption, a thin layer of water is formed on the surface of the substrate, which affects the propagation of surface acoustic waves. This effect is manifested in a decrease in the SAW velocity, therefore, in a change in the delay time. The thickness of the adsorption layer increases with a decrease in temperature, and when the temperatures of distilled water and the substrate are equal, condensation begins on the surface of the substrate. At the same time as the temperature decreases, the SAW velocity in the substrate material increases and the linear dimensions of the acoustic line decrease. The complex of these interconnected, temperature dependent parameters results in the nature of the change in SAW velocity and delay time being significantly complicated.

A method based on interference of the balanced antiphase signals was used to detect the change in the SAW velocity: the direct signal supplied from the generator and the signal passed by the SAW delay line.

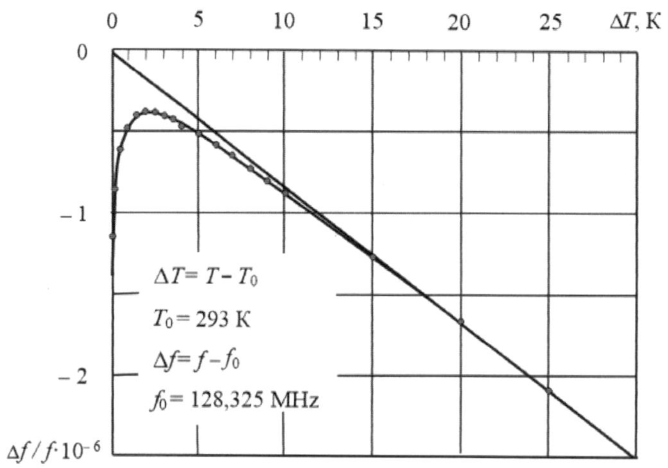

Figure 1. Temperature dependence of interference minimum frequency variation. *1* – *YZ*-cut substrate LiNbO3 (no adsorption layer), *2* – lithium niobate layered system – adsorbed water.

Temperature dependence of interference minimum frequency change in case of free surface of acoustic line is linear (Fig.1, curve 1). The presence of the adsorption layer complicates the nature of this relationship (curve 2), since when cooling the acoustic line, the change in delay time is influenced by two competing processes. First, the delay time is reduced as a result of the thermal increase in the SAW velocity and the reduction in the length of the acoustic line. Second, the delay time is increased due to a decrease in the SAW velocity caused by an increase in the thickness of the adsorption layer with a decrease in the temperature of the substrate (acoustic line).

In the first step of lowering the substrate temperature, the effect of the process of reducing the delay time (as a result of increasing the velocity) prevails (curve 2). When cooling a substrate thickness of an adsorptive layer increases, respectively, SAW velocity decreases and at some temperature frequency change of an interferential minimum reaches the maximum value. Obviously, in this case, the competing processes are balanced. Further, in the second step of cooling the substrate, the effect of the adsorption layer is increased by so much that the process of increasing the delay time (as a result of decreasing the velocity) is dominated.

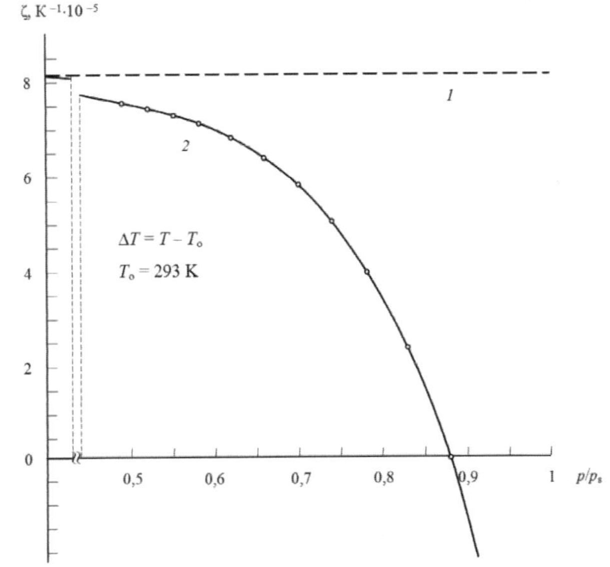

Figure 2. Temperature dependence of TCD of Rayleigh waves
1 – in dry gas environment; *2* in a humid gas environment.

On the basis of experimental data of temperature dependence of frequency change of an interferential minimum from a condition of equality it is possible to carry out the assessment of change of TCD of the Rayleigh waves in a layered system an adsorbed water – lithium niobate. On the basis of experimental data of temperature dependence of frequency change of an interferential minimum from a condition of equality $-\Delta f / f = \Delta\tau/\tau$ it is possible to carry out the assessment of change of TCD of the Rayleigh waves in a layered system an adsorbed water – lithium niobate. To do this, it is enough to take the temperature derivative of the function describing curve 2 (Fig. 1). From the experimental curve 2 (Fig. 1), it can be seen that with the rise in temperature of the TCD of the Rayleigh waves increases from negative to positive value and tends to the value of the TCD of the Rayleigh waves in the dry gas medium. Temperature of null value of TCD matches temperature of a maximum of a curve of frequency change of an interferential minimum.

Using the known relationship of relative vapor pressure to temperature (e.g., the Claveyron-Clausius equation), one can consider how the humidity of the gas medium affects the TCD of the Rayleigh waves in a layered system. The results of the analysis are shown in Figure 2. With increase of relative humidity TCD system decreases from positive to negative value. At relative humidity ~ 88% TCD takes zero value (Fig. 2, curve 2).

Thus, it is shown that the temperature coefficient delay time of the Rayleigh waves depends on both the degree of humidity and the temperature of the acoustic line. When humidity increases, TCD varies from positive to negative value. The largest change is observed at relative humidity values close to 100%. Destabilizing effect of ambient temperature and humidity must be taken into account when designing SAW- devices.

References

[1] Matthews H 1977 Surface wave filters. Design, construction, and use. Wiley-Interscience 521 p
[2] Simakov I G and Gulgenov Ch Zh 2009 *Tyumen state university bulletin* **6** 52–9.
[3] Simakov I G and Gulgenov Ch Zh 2011 *Buryat state university bulletin* **3** 216–20.

IOP Conf. Series: Materials Science and Engineering **704** (2019) 012007 doi:10.1088/1757-899X/704/1/012007

Preparation and Properties of Nanocomposites in the Systems S−AgI and S−Ag_2S−AgI in Dimethyl Sulfoxide

N V Khan *[1], M M Burkitbayev [1] and F Kh Urakaev *[1,2]
[1] Al-Farabi Kazakh National University, al-Farabi av. 71, Almaty 050040, Kazakhstan
[2] Sobolev Institute of Geology and Mineralogy, Siberian Branch, Russian Academy of Sciences, acad. Koptyug av. 3, Novosibirsk 630090, Russia

*) Emails: natasha.khan.v@gmail.com; urakaev@igm.nsc.ru

Abstract. In this research work described synthesis of nanocomposites in the systems S−AgI and S−Ag_2S−AgI in dimethyl sulfoxide (**DMSO**, $(CH_3)_2SO$) medium, and its some properties. Nanocomposites were obtained through the solvothermal synthesis at temperature range (**T**) 25÷160°C. The initial solutions were prepared by dissolution of sulfur (**S**), silver nitrate (**$AgNO_3$**) and ammonium iodine (**NH_4I**) in DMSO. The final products were obtained through the simple exchange reactions between initial solutions. Using the X-ray phase analysis (**XRD**), energy dispersive X-ray analysis (**EDAX**) and scanning electron microscopy (**SEM**) the samples were tested on phase and elemental composition, particle size and morphology, respectively. The results of XRD revealed that nanocomposites represented by S and AgI for system S−AgI, and by S, AgI and Ag_2S for S−Ag_2S−AgI system. The EDAX analysis showed the presence in composition of the investigated samples such elements like S, Ag and I for both systems. The SEM images demonstrated that both systems have a heterogeneus structure, the particles represented by flat and irregular shape. The size of the particles for S−AgI is fluctuated from 100 to 50000 nm and for S−Ag_2S−AgI – from 90 to 4000 nm.

1. Introduction

Nanotechnology is an integral part of modern science. Production and implementation of nanoparticles (***NPs***) and nanocomposites (***NCs***) in different branches of science and technology is commonplace. Such a great interest in *NPs* is due to the fact that particles in the state of nanoscale acquire properties that cannot be manifested in their usual state (routine size). Being in the state of nanosize, particles show antimicrobial, electronic, magnetic and strength characteristics [1]. With regard to the use of inorganic nanoparticles, the range of applications varies from biomedicine to electronics [1, 2]. The choice of the S−AgI and S−Ag_2S−AgI inorganic system is explained by the fact that each of its components has a set of specific properties. This in turn can lead to the production of *NCs* with enhanced multifunctional characteristics.

Application of each component of our system has a wide range. S is one of the most abundant elements in nature with semiconductor properties and high biological activity. Thanks to this features S, it allotropic and nano state, widely applied in agriculture, electronics, biochemistry, medicine [3–5]. Ag−*NPs* are attractive for various fields of science and technique, due to getting of novel and improved properties in comparison with bulk Ag [6]. For example, Ag−*NPs* has such features, like electrical conductivity, chemical stability, catalytic activity, photocatalytic activity and antimicrobial properties [6, 7]. Nanocrystaline silver halides (**AgX**; X = Cl, Br, I) are semiconductors with indirect and direct band gap [8] and photosensitivity. Such properties give possibility to use AgX in photographic films and as photocatalysts [8–12]. However, as was mentioned at [12] the pure AgI nanostructures inclined to photoreduction under visible light irradiation if no sacrificial reagent is supplied. The resolution of this problem is obtaining of composites with AgI or modification of surface of bulk material, which lead to the increasing of stability of the substance [12]. Ag_2S bulk material and Ag_2S−*NPs* also attractive for different fields, due to its antimicrobial, semiconductor, fluorescent, photocatalytic and other properties [13–16].

As for the methods for obtaining *NCs*, today there are no literature sources for obtaining of the systems S–AgI and S–Ag_2S–AgI. However, there are many methods of production of the individual materials, *NPs* or quantum dots of each component of our system or in combination with other precursors. In [17] was used 3D TiO_2 as suitable bases for fabrication a visible-light-induced 3D AgI@TiO_2 nanophotocatalyst. This composition revealed good photocatalytic properties and stability. Another compound with the Ag–AgI/Al_2O_3 system was prepared by deposition-precipitation and photoreduction methods. The system exhibited ability to the degradation and mineralization of toxic persistent organic pollutants [18]. In research [19] was obtained AgI–Ag_2S heterostructure with enhanced photothermal and optical absorption properties; this heterostructure was more efficient than AgI and Ag_2S separate *NPs*. In [20] was developed AgCl–Ag_2S–*NCs* via a simple one pot hydrothermal route, with enhanced photocatalytic properties. The study [11] was devoted to the obtaining of the AgI–Ag_2S@C_3N_4–*NCs* with ability to degradation of aqueous pollutants. Also Ag– and Ag_2S–*NPs* were studied on denitrification in sediments [15]. In study [21] were synthesized Ag–*NPs* in solutions of native and soluble starches in DMSO medium. As the result, Ag–*NPs* showed good stability and DMSO solution of soluble starch was characterized by better reductive activity than the native starch solution.

As can be seen, the development of the synthesis technology of *NCs* based on AgI and Ag_2S has a considerable interest. Moreover, the composites of the composition S–AgI and S–Ag_2S–AgI have not yet been obtained. Therefore the aim of this study was preparation of the S–AgI– and S–Ag_2S–AgI–*NCs* in DMSO medium. Based on this aim the following tasks were set: development of the modes of the synthesis and study of the possibility of obtaining of *NCs* in DMSO. The synthesized *NCs* were studied by means of XRD, SEM and EDAX.

2. Materials and methods

2.1 Materials

$AgNO_3$ (>99.8%, Sigma Aldrich), and chemically pure (LaborPharma) NH_4I, S and DMSO were used. For the preparation of water solutions and washing of the samples we used water purified by purification system Smart2Pure (Thermo scientific).

2.2 Synthesis of S–AgI

At the first stage of the synthesis there was preparing of the initial solutions at room temperature. According to the calculations for obtaining of the AgI 1.4468 g of $AgNO_3$ was dispersed into 50 mL of DMSO and stirred until full dissolution. The same operations were made with the 2.4680 g of NH_4I. The amount of NH_4I was doubled to avoid the formation of Ag_2S. When the initial solutions are ready, there are goes the obtaining of S solution in DMSO. The process occurs with help of installation which consist of three-neck-flask, backflow condenser, thermometer, silicone oil bath and magnetic stirrer. 100 mL of DMSO surged into the three-neck-flask and then 1 g of S is added. When all components were added, the system was exposed to the constant stirring (400 rpm) and gradual heating. S started dissolve at 110°C and at 120°C totally goes into the DMSO solution. When the mark T reached 120°C the heating is stopped. When the T of the reactional mixture reached ~50°C there are goes alternately adding of initial solutions of $AgNO_3$ and NH_4I. As soon as all the solutions were added, 150 mL of ultrapure water was added to the mixture in order to extract the S. Then the product of the synthesis was precipitate on centrifuge ROTINA380R for 15 minutes and at a speed 4000 rpm and washed by ultrapure water 2 times. Eventually the samples had been dried at 50°C, for 15-20 hours.

2.3 Synthesis of S-Ag_2S-AgI

As at the first method, which was described at previous paragraph, solutions of $AgNO_3$ and NH_4I were prepared in DMSO at room temperature. According to the stoichiometric calculations 1.0282 g of $AgNO_3$ was dissolved in 50 mL of DMSO and 0.3532 g of NH_4I in 50 mL of DMSO. When the initial solutions are ready, there are goes the obtaining of S solution in DMSO. In this method there was used the same installation which was mentioned above. All operations repeated like at first method, but the reactional system heated gradually to 160°C. When T reached 160°C there are goes alternately adding of initial solutions $AgNO_3$ and NH_4I. When initial solutions were added, the reactional system was cooled at room T during 12 hours. The finishing stage involved the precipitation of sediment and drying at previous paragraph.

2.4. Instruments

For the structural characterization of the investigated samples was a conducted XRD. The XRD of the samples was carried out using X-ray difractometer Rigaku MiniFlex 600 (CuKα radiation, λ = 1.54056 nm). Identification of the sample was made using the JCPDS–PDF2 database. For determination of the S was used literature source [22]. The SEM-analysis was conducted with help of scanning electron microscope Quanta 200i 3D, equipped by EDAX for investigation of the particle size and its morphology and elemental composition. The microscope was operated at an accelerating voltage 15 kV, a conductive carbon adhesive tape served as substrate for the samples.

3. Results and discussions

According to the XRD analysis the S-AgI nanocomposite is consist of S (S_8 form) and hexagonal β–AgI (figure 1). The black line is the experimental sample of the S–AgI–*NCs*. The red columns is S and green columns is AgI. As can be seen from the XRD patterns the intensive peaks values at 2 theta (**2θ**) = 22.30, 23.70, 25.35, 32.75, 39.20, 42.65 and 46.35° correspond to the reflections (100), (002), (101), (102), (110), (103) and (112) of β–AgI (JCPDS file no. 09-0374), respectively. Weak peaks 52.00, 59.30, 71.05 and 73.40° also correspond to the reflections (202), (203), (210), (300), (213) and (302) of β–AgI (JCPDS file no. 09-0374), respectively. More noticeable high peak for S_8 [22] is 23.10° and low peaks are 11.45, 15.40, 26.70, 27.75, 28.65, 31.40, 34.25, 34.90, 35.90, 37.05, 37.80 and 56.70°. Those peaks can be indexed to the reflections (222) and (111), (113), (311), (313), (044), (400), (333), (244), (317), (422) and (624), accordingly.

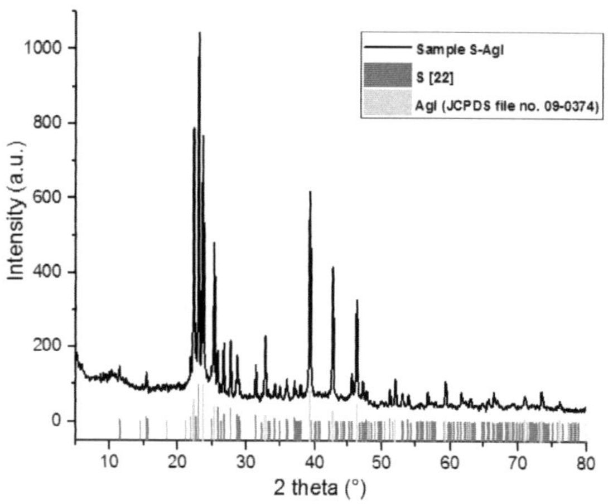

Figure 1. XRD patterns of the S–AgI

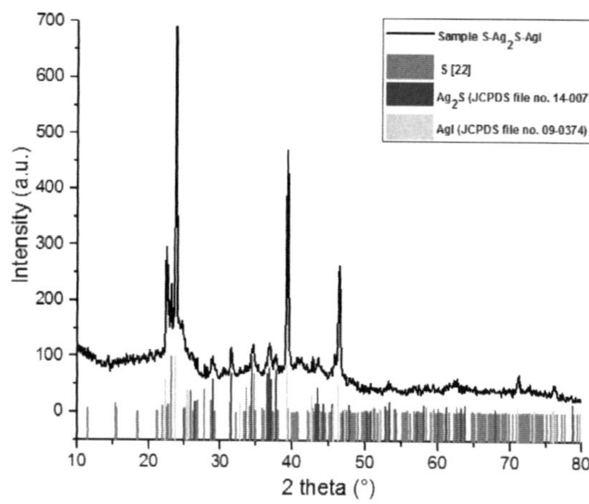

Figure 2. XRD patterns of the S–Ag_2S–AgI

The crystalline size was calculated using Debye-Scherrer equation [23]. According to the calculations the average crystallite size of S–AgI–*NCs* sample is about 140 nm.

Figure 2 shows the XRD patterns for S–Ag_2S–AgI. According to the results of analysis it can be seen that the sample has a characteristic peaks of S_8, mainly monoclinic α–Ag_2S (acanthite) and β–AgI. The black line corresponds to the experimental results, the red columns is S, the blue is Ag_2S and the green is AgI. The most distinct peaks for AgI are 22.40, 23.70, 32.65, 39.25, 42.75, 46.35, 71.15 and 76.05° correspond to the reflections (100), (002), (102), (110), (112), (300) and (302) of the β–phase, respectively. The Ag_2S (JCPDS file no. 14-0072) exhibit the weak diffraction peaks nearly 29.00, 31.50, 34.50, 36.81, 37.70 and 43.45°, which were indexed as (111), (–112), (–121), (121), (–103) and (200) diffraction plane, respectively. The appearance of such peaks like 23.10 and 27.75° is attributed by (222) and (206) reflections of the S_8 [22]. The crystalline size of the S–Ag_2S–AgI–*NCs* was calculated and the average size was nearly 80 nm.

IOP Conf. Series: Materials Science and Engineering **704** (2019) 012007 doi:10.1088/1757-899X/704/1/012007

EDAX analysis confirmed the presence of elements, which were founded by XRD for both systems. According to EDAX the S–AgI sample consists of S, Ag and I (figure 3ab and table 1).

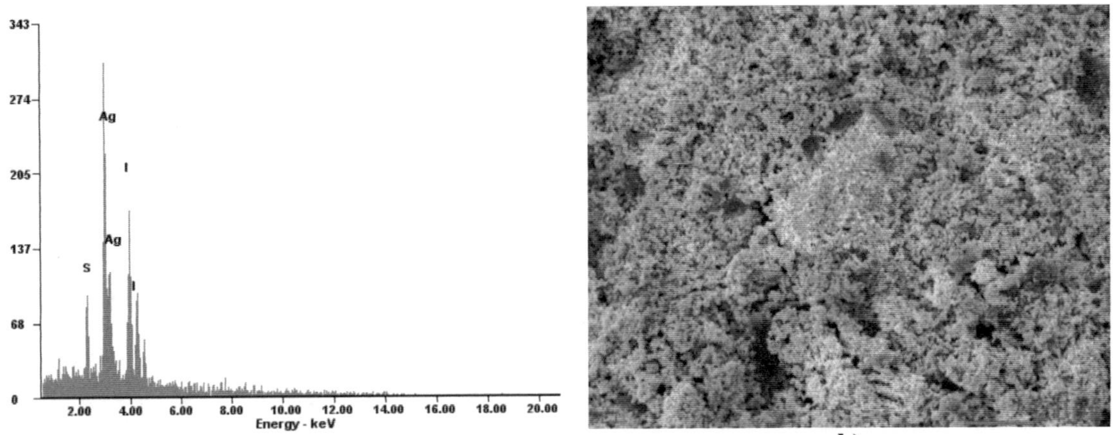

a) b)

Figure 3. EDAX of the S–AgI: **a**) the spectra; **b**) the area from which the analysis data was obtained

Table 1. The percentage data of the EDAX analysis of the S–AgI

Element	Wt%	At%
S	4.08	13.43
Ag	45.81	44.86
I	50.11	41.71

Pointed investigation shows that sample represented by light particles (in basis) and also there can be seen the dark flat particles of big sizes (figure 4ab). Elemental composition of that area (figure 4a and table 2) revealed that the dark particle is consist of S on 87.42 At%, remaining 6.44 and 6.15 At% are Ag and I (the light small particles, which settled on the surface of the S (figure 4b), respectively.

a) b)

Figure 4. EDAX of the S–AgI: **a**) the spectra; **b**) the area from which the analysis data was obtained

Table 2. The percentage data of the EDAX analysis of the dark area of the S–AgI

Element	Wt%	At%
S	65.54	87.42
Ag	16.23	6.44
I	18.23	6.15

Elemental composition of the S–Ag_2S–AgI in general represented by S (13.31 At%), Ag (45.38 At%) and I (22.35 At%), see figure 5 and table 4.

a) **b**)

Figure 5. EDAX of the S–Ag_2S–AgI: **a**) the spectra; **b**) the area from which the data was obtained

Table 3. The percentage data of the EDAX analysis of the S-Ag_2S-AgI

Element	Wt%	At%
O	3.59	18.97
S	5.04	13.31
Ag	57.85	45.38
I	33.52	22.35

As can be seen on the figure 6b the S–Ag_2S–AgI sample also presented in some areas by dark and flat particles. The analysis of that zone proved that the particle consists of 91.90 At% of S (table 5).

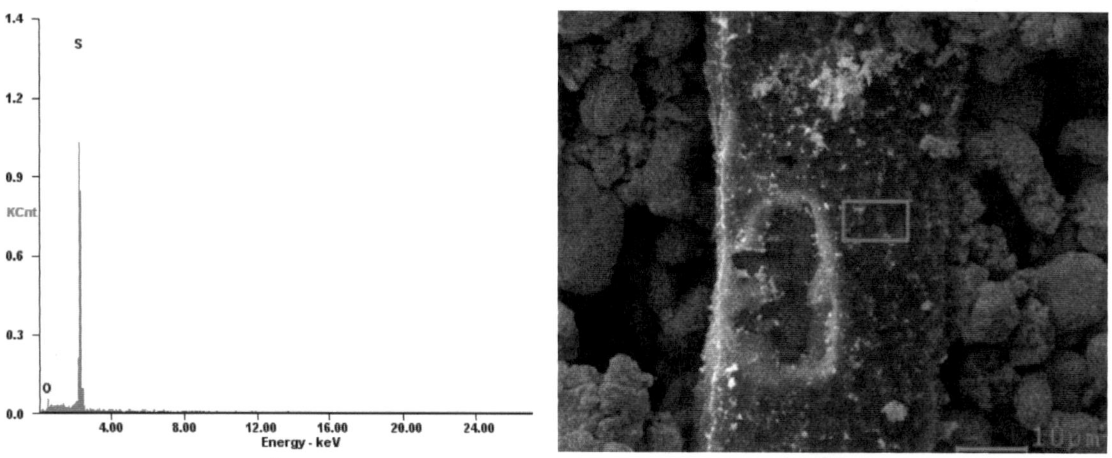

a) **b**)

Figure 6. EDAX of the S–Ag_2S–AgI (black area): **a**) the spectra; **b**) the area from which the data was obtained

Table 4. The percentage data of the EDAX analysis of the S-Ag_2S-AgI (black area)

Element	Wt%	At%
O	4.21	8.10
S	95.79	91.90

After phase investigation there was made characterization of morphology of the samples and study of its particle size. The images of the S–AgI sample are shown on the figure 7.

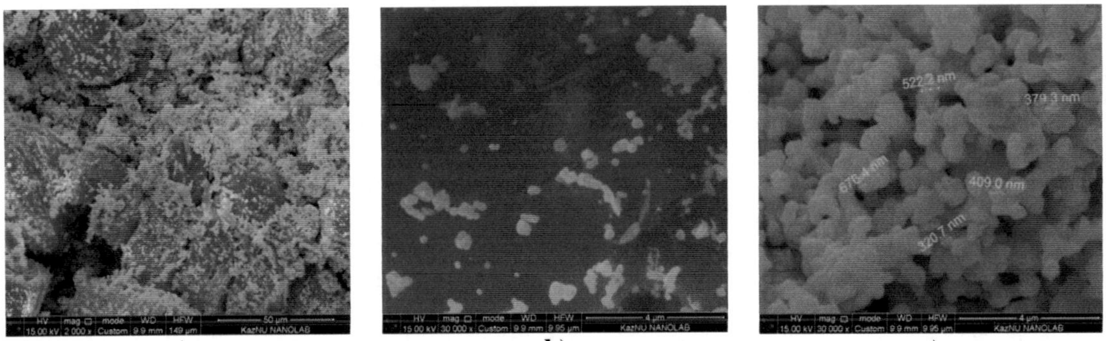

a) **b)** **c)**

Figure 7. SEM-analysis of the S–AgI (**a**, **b**, **c**)

According to the SEM the sample is heterogeneous; the particles are represented by both large and small sizes (figure 7a). The large particles of dark colour has the size about 2-5 μm. More detail study of the dark particles (figure 7b) revealed that on its surface settled the light particles of AgI (100-1000 nm) according to EDAX. The size of the S–AgI particles fluctuated from ~100 nm to 5 μm (figure 7abc). As for the morphology of the sample, it is obvious that the dark particles of S (figure 4b) are flat and light particles are mainly oval or sphere shape with a smooth surface.

The images of the S–Ag_2S–AgI sample are shown on the figure 8.

a) b) c)

Figure 8. SEM-analysis of the S–Ag_2S–AgI obtained in DMSO medium (**a**, **b**, **c**)

The S–Ag_2S–AgI also has non-uniform mass (figure 8ab). There are also present light (figure 8a) and dark (figure 8c) particles. The size of the particles varies from ~90 nm to 4 μm. In general particle has irregular shape.

In both cases there is a strong defect of crystallites and coalesce of small crystallites, which complicates the determination of real sizes of particles. These syntheses were carried out at high temperatures, which make it hard to control the size of the resulting particles.

Also, the complex composition of the system leads to the difficulty of controlling the particle size, since each of the constituent components has own "behavioral" features. In other hand, according to the XRD and SEM analyses, both methods can give nanosized particles; the problem is that small crystallites coalesce with each other and form large agglomerates. Stabilization of the particles can serve as solution of the mentioned problem.

In addition, the use of DMSO as a solvent makes it possible to apply a wide temperature range for synthesis. Thus, it can allow not to be tied to one temperature regime. DMSO is an aprotic bipolar solvent. It is widely used in medicine, biology and pharmacy [24]. Therefore, there is no need to be afraid of using this solvent from an environmental point of view.

4. Conclusion

In this study, the synthesis of S-AgI and S–Ag_2S–AgI nanocomposites was carried out for the first time. The synthesis passed with using precursor solutions in dimethyl sulfoxide, at temperatures above 120°C. By XRD it was established that obtained nanocomposites have characteristic diffraction peaks of S in S_8 form and β–AgI for S–AgI, and S_8, α–Ag_2S and β–AgI for S–Ag_2S–AgI.

Owing to the EDAX analysis was confirmed the presence of S, Ag and I in both cases, and the pointed EDAX showed that dark flat particles represented by S. The EDAX spectra of S–Ag_2S–AgI nanocomposite revealed the presence of the O, this indicates impurity of obtained crystallites.

The XRD also showed that the S–AgI has an average size of the particles equal to 140 nm, meanwhile S–Ag_2S–AgI – 80 nm. However, SEM analysis did not confirm this data. According to the SEM the size of the S–AgI particles is about from 100 nm 5 μm and S–Ag_2S–AgI particles is about from 90 nm to 4 μm. An attempt was made to explain the differences in the results of these methods.

Acknowledgments

This research work was supported by scientific projects of the Ministry of Education and Science of the Republic of Kazakhstan (AP05133115 and program-target financing BR05234566), RFBR grants (18-05-00434A and 19-45-540003p_a) and is done on state assignment of IGM SB RAS.

References

[1] Stark W J, Stoessel P R, Wohlleben W and Hafner A 2015 Industrial application of nanoparticles *Journal the Royal Society of Chemistry* **44** (16) 5793–805 DOI: 10.1039/c4cs00362d

[2] Lu A H, Salabas E L and Schüth 2007 Magnetic nanoparticles: synthesis, protection, funtionalization, and application *Angewandte Chemie International Edition* **46** (8) 1222–44 DOI: 10.1002/anie.200602866

[3] Griffith C M, Woodrow J E and Seiber J N 2014 Environmental behavior and analysis of agricultural sulfur *Pest management science* **71** (11) 1486–96 DOI 10.1002/ps.4067

[4] Teng Y, Zhou Q and Gao P 2019 Applications and challenges of elemental sulfur, nanosulfur, polymeric sulfur, sulfur composites, and plasmonic nanostructures *Critical Reviews in Environmental Science and Technology* pp 1–45 https://doi.org/10.1080/10643389.2019.1609856

[5] Xiao P, Sun L, Liao D, Agboola P O, Shakir I and Xu Y 2018 Unexpected effects of electrode architecture on high-performance lithium-sulfur batteries *ACS applied materials & interfaces* **10** (39) 33269–75 DOI: 10.1021/acsami.8b11883

[6] Haider A and Kang I-K 2015 Preparation of the silver nanoparticles and their industrial and biomedical applications: a comprehensive review *Advances in Materials Science and Engineering* **2015** http://dx.doi.org/10.1155/2015/165257

[7] Mohler J S, Slim W, Blaskovich M A T, Cooper M A and Ziora Z M 2016 Silver bullets: A new luster on an old antimicrobial agent *Biotechnology advances* **36** 1391–411

[8] Abeyweera S C, Rasamani K D and Sun Y 2017 Ternary silver halide nanocrystals *Accounts of chemical research* **50** (7) 1754–61 https://doi.org/10.1021/acs.accounts.7b00194

[9] Marucci G, Monno A and van der Werf I D 2014 Non invasive micro-Raman spectroscopy for investigation of historical silver salt gelatin photographs *Microchemical Journal* **117** 220–4 http://dx.doi.org/10.1016/j.microc.2014.07.001

[10] Su M, Liu H and Ma J 2016 Synthesis of cube like Ag@AgCl photocatalyst with enhanced photocatalytic activity *Journal of Material Science: Materials in electronics* **27** (10) 10707-11 DOI: 10.1007/s10854-016-5171-4

[11] Velmurugan S, Balu S, Palanisamy S, Yang T C-K, Velusamy, Chen S-W and El-Shafey E-S I 2019 Synthesis of novel environmental sustainable AgI-Ag_2S nanospheres impregnated g-C_3N_4 photocatalyst for efficient degradation of aqueous pollutants *Applied surface science* S0169–4332(19)32807–7 https://doi.org/10.1016/j.apsusc.2019.143991

[12] Reddy D A, Choi J, Lee S, Ma R and Kim T K 2015 Green synthesis of AgI nanoparticle-functionalized reduced grapheme oxide aerogels with enhanced catalytic performance and facile recycling *RSC Advances* **83** 67394–404 DOI: 10.1039/C5RA07267K

[13] Liu Y, Yang T, Wang L, Huang Zh, Li J, Cheng H, Jiang J, Pang S, Qi J and Ma J 2018 Interpreting the effects of natural organic matter on antimicrobial activity of Ag_2S nanoparticles with soft particle theory *Water research* **145** pp 12–20 DOI: 10.1016/j.watres.2018.07.063

[14] Ayodhya D and Veerabhadram G 2016 Green synthesis, characterization, photocatalytic, fluorescence and antimicrobial activities of *Coclosperum gossypium* capped Ag_2S nanoparticles *Journal of Photochemistry and photobiology* **157** pp 57–69 DOI: 10.1016/j.jphotobiol.2016.02.002

[15] Liu S, Wang Ch, Hou J, Wang P, Miao L, Fan X, You G and Xu Y 2018 Effects of Ag and Ag_2S nanoparticles on denitrification in sediments *Water research* **137** 28–36 https://doi.org/10.1016/j.watres.2018.02.067

[16] Xue J, Liu J, Mao S, Wang Y, Shen W, Wang W, Huang L, Li H and Tang J 2018 Recent progress in synthetic methods and applications in solar cells of Ag_2S quantum dots *Material Research Bulletin* **106** 113–23 https://doi.org/10.1016/j.materresbull.2018.05.041

[17] An C, Jiang W, Wang J, Wang S, Ma Z and Li Y 2013 Synthesis of three-dimensional AgI@TiO_2 nanoparticles with improved photocatalytic performance *Dalton Transactions* **42** 8796–801 DOI 10.1039/c3dt50736j

[18] Hu Ch, Peng T, Hu X, Nie Y, Zhou X, Qu J and He H 2010 Plasmon photodegradation of toxic pollutants with Ag-AgI/Al_2O_3 under visible light irradiation *Journal of the American Chemical Society* **132** (2) 857–62 https://doi.org/10.1021/ja907792d

[19] Zeng W, Suo L, Zhang C, Wu D and Zhu H 2019 AgI–Ag_2S heterostructures for photothermal conversion and solar energy harvesting *Journal of the Taiwan Institute of Chemical Engineers* **95** 273–80 https://doi.org/10.1016/j.jtice.2018.07.012

[20] Zarghami Z, Ebadi M, Motevalli K and Alibadi M 2016 Self-assembly of nanoparticles to prepare of dentritic AgCl-Ag_2S nanocomposites through a simple one-pot hydrothermal approach and its photocatalytic activity *Journal of Material Science: Materials in Electronics* **27** (2) 1087–91 DOI 10.1007/s10854-015-3855-9

[21] Kochkina N E, Skobeleva O A and Khokhlova Yu V 2015 Synthesis of silver nanoparticles in DMSO solutions of starch: a comparative investigation of native and soluble starches *Nanosystems: physics, chemistry and mathematics* **6** (3) 405–11 DOI 10.17586/2220-8054-2015-6-3-405-411

[22] Coppens P, Yang Y W, Blessing R H, Copper W F, and Larsen F K 1977 The experimental charge distribution in sulfur containing molecules. Analysis of cyclic octasulfur at 300 and 100 K *Journal of the American Society* **99** (3) 760–6 https://doi.org/10.1021/ja00445a017

[23] Murugesan P, Narayanan S, Matheswaran M, Praveen Kumar M and Ravichandran S A direct Z-Scheme plasmonic AgCl@g-C_3N_4 heterojunction photocatalyst with superior visible light CO_2 reduction in aqueous medium 2018 *Applied Surface Science* S0169–4332(18)31073-0 https://doi.org/10.1021/acsami.5b09901

[24] Vignes R 2000 Dimethyl sulfoxide (DMSO) – A "New" clean, unique, superior solvent *American chemical society. Annual meeting* **228** pp 1–20 https://pdfs.semanticscholar.org/02a7/0304ac166372de26ca50077e6d3131786a3a.pdf

Simulation of femtosecond pulse in a Kerr-lens mode locked Ti: sapphire laser

E Nomin-Erdene*, D Unurbileg, P Munkhbaatar, Ts Baatarchuluun, Ts Khos-Ochir, G Munkhbayar, N Tuvjargal, G Erdene-Ochir, J Davaasambuu

Laser Research Center, National University of Mongolia, 14201, Ulaanbaatar, Mongolia

e-mail: nominerdene@num.edu.mn

Abstract. The Kerr-lens mode-locking (KLM) is known as a suitable method for generation of femtosecond pulses and mode-locked Ti:sapphire laser is now widely used sources of stable, energetic femtosecond pulses. We will present the simulation of KLM in Ti:sapphire laser cavities with a folded-cavity four-mirror by applying the *ABCD* ray-tracing technique for a Gaussian beam. Simulations will be performed for an asymmetric resonator design. Based on the numerical analysis, we will find the optimum design parameters (slit position, gain cavity spacing, gain medium position) for KLM.

Introduction

The Kerr-lens mode-locking (KLM) based on the lens-effect induced in suitable material by a Kerr nonlinearity is straight forward method to generate pulses as short as few femtosecond laser pulses [1]. When a beam propagates through a medium which a refractive index linearly varies with the intensity (Kerr nonlinearity), the well-known optical phenomenon that a self-focusing effect is observed [2]. The self-focusing effect of an intense pulses is successfully exploited in the mode-locking [3]. The optical Kerr effect is responsible for the intensity dependent refractive index change which causes the lensing in the gain medium. When a suitable aperture is located inside the cavity where the intense pulses beam profile is narrowed by the Kerr-lens effect, the continuous-wave (CW) radiation propagation losses are higher than the intense pulse propagation losses [4]. The combination of the Kerr-lens and aperture acts as a fast saturable absorber, then a fast passive gain modulation is obtained [5]. The intensity fluctuations in laser start-up are not intense enough to induces Kerr lensing [6,7], thus acousto-optic modulation, additive pulse mode-locking, impulsive starting i.e [8-17] are required before the KLM. The self-starting KLM is achieved by using a highly nonlinear Kerr medium in the folded cavity.

In this paper we simulate the influence of the Kerr lensing on the beam propagation in a folded-cavity four-mirror resonator by applying the ABCD ray-tracing technique for a Gaussian beam. With a split-step method that dividing the Ti:sapphire crystal to many thin lens-like slices [18] the intensity dependent Kerr-lens effect of the Brewster-plate shaped active medium is described. We introduce the optimum design parameters and discuss the modifications of the beam spot size and radius of curvature in the nonlinear Kerr medium, where the refractive index increases with intensity.

Beam propagation in the resonator

The folded-cavity four-mirror resonator, shown in Fig.1 is used for Ti:sapphire lasers. Parameters l_{12}, l_{23}, l_{34} are lengths of the three arms and R_2, R_3 are curvature radii of M_2 and M_3 mirrors. The beam propagation inside the cavity is expressed by the ABCD round-trip matrices. Due to the astigmatic effect the beam propagation behaviour is different in the tangential xz plane and in the sagittal yz plane [19]. The tangential and sagittal rays focus at different positions (astigmatism). This cavity astigmatism is compensated by the Brewster plate at a certain tilting angle. Without the Kerr lensing the Brewster plate behaves linearly, i.e. ($n_2=0$) [3]. In the case with Kerr lensing the intensity dependence of the refractive index is taken into account in the beam propagation through the gain medium.

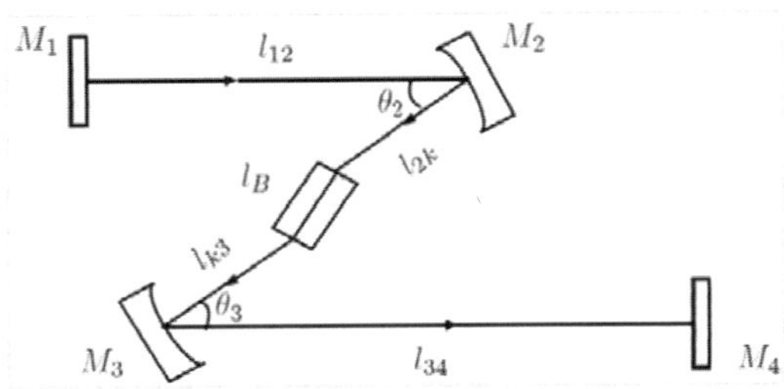

Figure 1. The folded-cavity four-mirror resonator with the Brewster-plate shaped active medium.

For one round trip through the resonator the total ABCD matrix which is the multiplication of all the matrices of the elements is given by

$$M_t = \begin{pmatrix} A_t & B_t \\ C_t & D_t \end{pmatrix} = \prod_i M_i = \prod_i \begin{pmatrix} A_t & B_t \\ C_t & D_t \end{pmatrix}$$
$$= M_1 M_{12} M_2 M_{2k} M_k M_{k3} M_3 M_{34} M_4 M_{34} M_3 M_{k3} M_k M_{2k} M_2 M_{12}$$

The optical elements in the resonator are listed in Table I. To be a stable resonator, a stability condition, $|A_t + D_t| \leq 2$ [20] is required which gives the allowed regions of mirror separation $l_{23} = l_{2k} + l_B + l_{k3}$ for stable laser operation. In this stability region the resonator misalignment sensitivity parameter is determined by $|1/C_t|$ where small value of $|1/C_t|$ implies lower sensitivity of the resonator [5].

Using the ABCD matrices one can calculate the fundamental Gaussian TEM_{00} mode of the cavity for CW operation inside the stability zones of the resonator. The Gaussian mode is represented by the complex beam parameter as

$$\frac{1}{q(z)} = \frac{1}{R(z)} - i\frac{\lambda_L}{\pi w^2(z)} \tag{1}$$

where $R(z)$ and $w(z)$ are the wavefront curvature and the spot size at position z, λ_L is the wavelength. The beam diameter and spot size are related by $\Delta d = [2\ln 2]^{1/2} w$. The beam parameter $q(z)$ for the propagation direction z from z_1 to z_2 is given by

$$q(z_2) = \frac{A_{12} q(z_1) + B_{12}}{C_{12} q(z_1) + D_{12}} \tag{2}$$

The incident beam has the same curvature as mirrors. In our resonator the input mirror M_1 at $z = 0$ is plane, with the radius $R_1(0) = \infty$. Then, the beam parameter for one round trip on this mirror is found as

$q[0] = i\pi w^2(0)/\lambda_L = [A_t q(0) + B_t]/[C_t q(0) + D_t]$. Without the Kerr lensing the ABCD matrix components in the equation are constant, thus this quadratic equation can be solved for $q(0)$ and $w(0)$. Considering the Kerr-lens effect, the ABCD matrix components depend on the beam spot size w [3]. To start the ray tracing in the resonator an arbitrary beam parameter is chosen and the calculation is repeated until the real part of the beam parameter $q_{m+1}(0)$ and $q_m(0)$ at the M_1 mirror equals to zero for one round trip.

$$q_{m+1}(0) = \frac{i\pi w_{m+1}^2(0)}{\lambda_L} = \frac{[A_t q_m(0) + B_t]}{[C_t q_m(0) + D_t]} \tag{3}$$

The intensity changes inside the Kerr lens medium is achieved by dividing the crystal to many thin lens-like slices ($l_B \rightarrow 0$) and the beam propagation through each slices is calculated by using the Kerr-lens ABCD matrices given in Table.I with a constant intensity I_{0L}. The solution depends on the crystal position [21].

Table I. ABCD matrices

Optical element	Matrix	
Mirror		
Normal incidence	$\begin{pmatrix} 1 & 0 \\ -2/R & 1 \end{pmatrix}$	M_1, M_4
Tilted, sagittal plane	$\begin{pmatrix} 1 & 0 \\ (-2\cos\vartheta)/R & 1 \end{pmatrix}$	M_2, M_3
tangential plane	$\begin{pmatrix} 1 & 0 \\ -2/(R\cos\vartheta) & 1 \end{pmatrix}$	M_2, M_3
Propagation in space	$\begin{pmatrix} 1 & l \\ 0 & 1 \end{pmatrix}$	$M_{12}, M_{2k}, M_{k3}, M_{34}$
Plate		
Normal incidence	$\begin{pmatrix} 1 & l/n_L \\ 0 & 1 \end{pmatrix}$	
Brewster tilted, sagittal plane	$\begin{pmatrix} 1 & l_B/n_L \\ 0 & 1 \end{pmatrix}$	M_k
tangential plane	$\begin{pmatrix} 1 & l_B/n_L^3 \\ 0 & 1 \end{pmatrix}$	M_k
Kerr-lens plate (tangential thickness l_B)		
Normal incidence	$\begin{pmatrix} \cos(\gamma l) & (l/\tilde{n}\gamma)\sin(\gamma l) \\ -\tilde{n}\gamma\sin(\gamma l) & \cos(\gamma l) \end{pmatrix}$	
Tilted to Brewster angle Sagittal plane	$\begin{pmatrix} \cos(\gamma_s l_B) & (l/\tilde{n}_B\gamma_s)\sin(\gamma_s l_B) \\ -\tilde{n}_B\gamma_s\sin(\gamma_s l_B) & \cos(\gamma_s l_B) \end{pmatrix}$	M_k
Tangential plane	$\begin{pmatrix} \cos(\gamma_t l_B) & (l/\tilde{n}_B\gamma_t n_L^2)\sin(\gamma_t l_B) \\ -\tilde{n}_B\gamma_t\sin(\gamma_t l_B) & \cos(\gamma_t l_B) \end{pmatrix}$	M_k

Table II. Resonator parameters (λ_L=800nm) [22]

Parameter	Value
Mirror curvatures (mm)	
R_1	∞
R_2	50
R_3	50
R_4	∞
Mirror tilting angles (deg)	
θ_1	0
θ_2	15.22
θ_3	15.22
θ_4	0
Total resonator length (m)	
l_t	1.05
Mirror separators (cm)	
l_{12}	*30*
$l_{23}=l_{2k}+l_B+l_{3k}$	varied around 50
l_{34} l_t-l_{12}-l_{23}	
Intracavity plate (Brewster-cut Ti:sapphire crystal)	
n_L	1.76
n_2	$1.6\times10^{-22} m^2V^{-2}$ *[22]*
l_B	2 mm
l_{2k}	varied around $l_{23}/2$

Changing the position of the crystal along the beam affects the mode intensity in the crystal. The optical Kerr effect is a third-order nonlinear process in which the refractive index of the material is intensity dependent [23]. At sufficiently high intensity, the refractive index of the medium will be influenced to a readily observable extent by the field intensity. The Kerr medium is described by the intensity dependent refractive index *n* that is given [1] by

$$n = n_L + \frac{1}{2}n_2 E_{0L}^2 = n_L + \gamma_2 I_L = n_L + \frac{n_2}{n_L c_0 \varepsilon_0} I_L \tag{4}$$

where n_2 and γ_2 are the nonlinear refractive index and nonlinear coefficient of medium. E_{0L}^2 is the electric field amplitude, $I_L = (n_L c_0 \varepsilon_0 / 2)E_{0L}^2$ is the light intensity c_0 being the light velocity in vacuum and ε_0 the electric permitivity of vacuum).

The Gaussian beam intensity is approximated by a Taylor expansion as

$$I_L = I_{0L}\exp[-2(r/w)^2] \approx I_{0L}[1-2(r/w)^2] \tag{5}$$

then, the (4) equation becomes parabolic [24].

$$\begin{aligned} n &= n_L + \frac{n_2 I_{0L}}{n_L c_0 \varepsilon_0}[1-2(r/w)^2] \\ &= (n_L + \frac{n_2 I_{0L}}{n_L c_0 \varepsilon_0})(1 - \frac{2n_2 I_{0L}}{n_L c_0 \varepsilon_0 (n_L + \frac{n_2 I_{0L}}{n_L c_0 \varepsilon_0})w^2} r^2) \\ &= \tilde{n}(1-\frac{1}{2}\gamma^2 r^2) \end{aligned} \tag{6}$$

where $\tilde{n} = n_L + \frac{n_2 I_{0L}}{n_L c_0 \varepsilon_0}$ and $\gamma = (\frac{4n_2 I_{0L}}{n_L c_0 \varepsilon_0 \tilde{n}})^{1/2}\frac{1}{w} = (\frac{8n_2 P}{n_L c_0 \varepsilon_0 \tilde{n}\pi})^{1/2}\frac{1}{w^2}$

The laser power is $P = \pi w^2 I_{0L}/2$. Since the Kerr lensing for the Brewster plate in the sagittal and tangential plane is different, the w and I_{0L} in (6) equation are replaced by $w \to w_t = n_L w$, $w \to w_s = w$, and $I_{0L} \to I_{0L} w^2/(w_t w_s) = I_{0L}/n_L$. The equations for the sagittal plane of a Brewster plate are given by $n_s = \tilde{n}_B(1-\frac{1}{2}\gamma_s^2 r^2)$ where $\tilde{n}_B = n_L + \frac{n_2 I_{0L}}{n_L^2 c_0 \varepsilon_0}$ and

$$\gamma_s = (\frac{4n_2 I_{0L}}{n_L^2 c_0 \varepsilon_0 \tilde{n}_B})^{1/2}\frac{1}{w} = (\frac{8n_2 P}{n_L^2 c_0 \varepsilon_0 \tilde{n}_B \pi})^{1/2}\frac{1}{w^2} \tag{7}$$

For the tangential plane of a Brewster plate the equation are written as

$$n_t = \tilde{n}_B(1-\frac{1}{2}\gamma_t^2 r^2)$$

$$\text{with } \gamma_t = (\frac{4n_2 I_{0L}}{n_L^4 c_0 \varepsilon_0 \tilde{n}_B})^{1/2}\frac{1}{w} = (\frac{8n_2 P}{n_L^4 c_0 \varepsilon_0 \tilde{n}_B \pi})^{1/2}\frac{1}{w^2} \tag{8}$$

The Kerr-lens ABCD matrices in the tangential and sagittal plane are given in Table I.

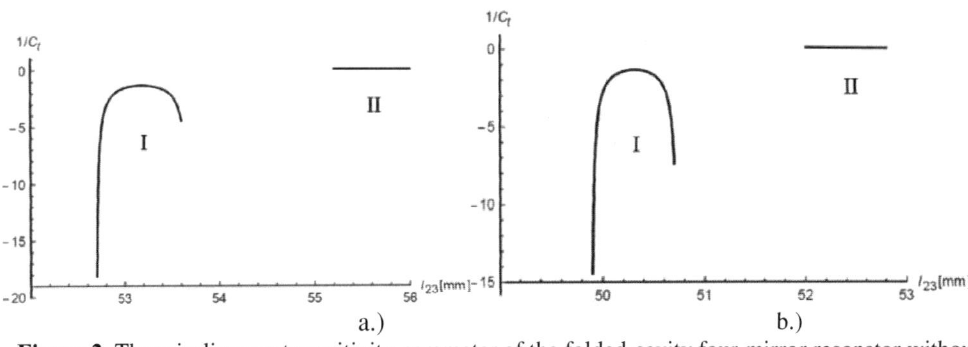

Figure 2. The misalignment sensitivity parameter of the folded-cavity four-mirror resonator without the Kerr lensing (P=0). a) Sagittal plane, b) tangential plane

Simulations

There are two stability regions observed in the tangential and sagittal plane of the four-mirror cavity with the astigmatism compensated. Due to the folded mirror separation l_{23} it is divided into two stability zones I and II as shown in the misalignment sensitivity parameter plot (Fig.2) without the Kerr lensing.

IOP Conf. Series: Materials Science and Engineering **704** (2019) 012008 doi:10.1088/1757-899X/704/1/012008

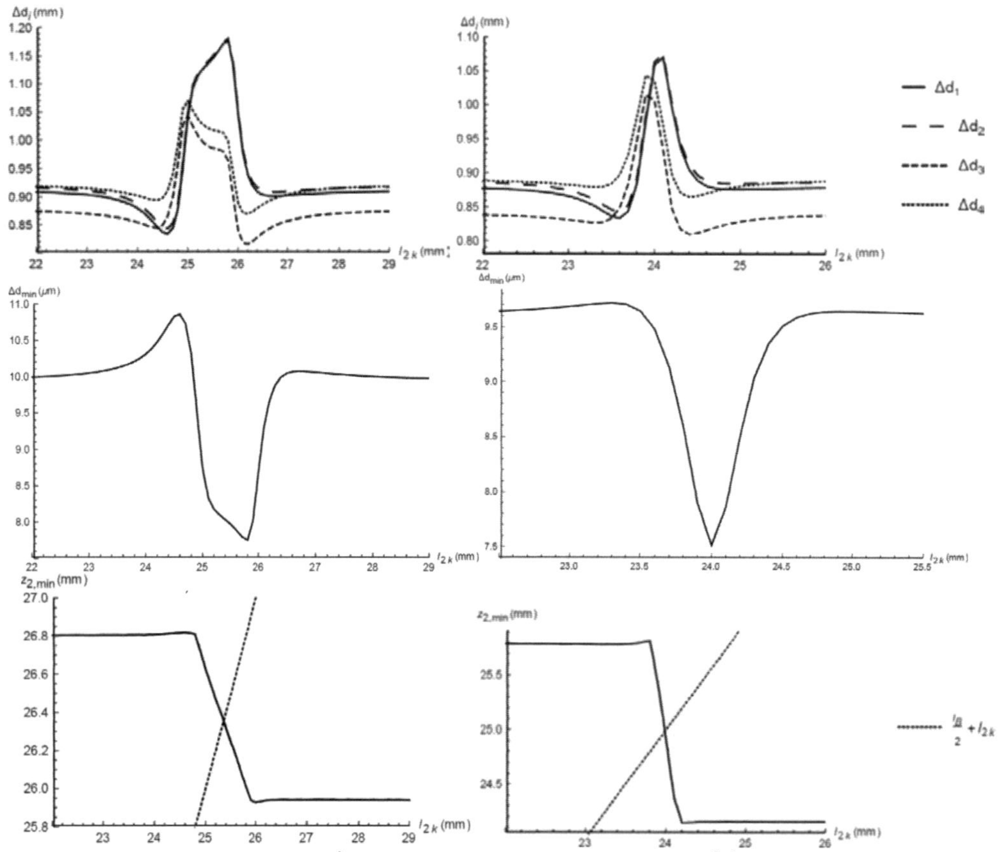

a.) b.)

Figure 3. The influence of the Kerr lens position l_{2k} position on beam parameters. Laser power, $P = 2\times10^5 W$. Parameters listed in Table II are applied. a.) sagittal plane, b.) tangential plane.

In the case with the Kerr lensing the parameter l_{23} is fixed and the Kerr lens position l_{2k} only in the folded cavity is varied. In Fig.3 the beam parameters modifications caused by the Kerr lensing are compared in the tangential and sagittal plane, respectively. The applied resonator parameters are listed in Table II. The plots, shown in Fig.3 corresponds to the beam diameters at the mirrors, the minimum beam diameters in the gain cavity and the position of the beam waist $Z_{2,\min}$ measured from the mirror M_2 against the Kerr lens position l_{2k}. The laser power inside the resonator is set to $P = 2\times 10^5$ W.

From the beam waist position plots, one can see that by increasing the distance from the M_2 to the crystal (for $l_{2k}>z_{2,min}$), the beam waist exists outside of the crystal, then its position remains constant. It is same for $l_{2k}<z_{2,min}-l_B$ and it shows the lensing effect getting small. When the beam waist exists inside the crystal ($l_{2k}<z_{2,min}< l_{2k} +l_B$), its position decreases until the end of the crystal because of the increase of the beam intensity inside the crystal. Since the beam intensity is high at the face of the crystal, around $l_{2k}=z_{2,min}-l_B$ and $l_{2k}=z_{2,min}$ the Kerr-lens effect is largest.
The beam size is changed in the Brewster plate, causing beam narrowing at some mirrors and beam broadening at others as shown in the plots with the beam diameters dw_i (Fig.3). An aperture setting in regions of beam narrowing leads to Kerr-lens mode locking.

Conclusion

We introduced the beam parameters modifications caused by the Kerr lensing in the four-mirror folded-cavity resonator. In the simulation the real parameters of a femtosecond Ti:sapphire laser are used. The astigmatism due to the beam propagation in the sagittal and tangential plane of the folded-cavity is compensated.

The optimum resonator design parameters (slit positioning, gain cavity spacing, gain medium position) are obtained for maximum Kerr-lens mode-locking effect.

Acknowledgements
This work is supported by the fundamental research project No.2019/05, financed by the Mongolian Foundation for Science and Technology.

References

[1] Svelto O 2010 *Principles of Lasers*, New York, Springer.
[2] Magni V and Cerullo G 1993 ABCD matrix analysis of propagation of gaussian beams through Kerr media, *Opt. Comm* **96** 348-55.
[3] Penzkofer A and Wittmann M 1996 Kerr lens effects in a folded-cavity four-mirror linear resonator, *Optical and Quantum Electronics* **28** 423-42.
[4] Spence D E and Kean P N 1991 60-fsec pulse generation from a self-mode-locked Ti:sapphire laser, *Opt. Lett* **16** 42.
[5] Malakzadeh A, Javad Kamjoo M 2017 Simulation of Kerr lens mode locking behavior in sagittal plane, *Laser Opt. Photonics* **4** 1.
[6] Chen S and Wang J 1991 Self-starting issues of passive self-focusing mode-locking. *Opt. Lett* **16** 1689-91.
[7] Curley P and Ferguson A 1991 Actively mode-locked Ti: sapphire laser producing transform-limited pulses of 150-fs duration, *Opt. Lett* **16** 1016-8.
[8] Spence D E and Evans J 1991 Regeneratively initiated selfmode-locked Ti: sapphire laser, *Opt Lett* **16** 1762-4.
[9] Sarukura N and Ishida Y 1991 Generation of 50-fsec pulses from a pulse-compressed, CW, passively mode-locked Ti: sapphire laser, *Opt Lett* **16** 153-5.
[10] French P, Kelly S 1989 Femtosecond pulse generation from a titanium-doped sapphire laser using nonlinear external cavity feedback, *Opt. Lett* **14** 686-8
[11] French P., Noske D. 1991 Characterization of a cw titanium-doped sapphire laser mode-locked with a linear external cavity, *Opt Comm* **83** 185-194.
[12] Pelouch W, Powers P. and Tang C. (1992) Ti: sapphire-pumped, high-repetition-rate femtosecond optical parametric oscillator, *Opt Lett* **17** 1070-2.
[13] Liu Y M, Sun K W and Prucnal P 1992 Simple method to start and maintain self-mode-locking of a Ti: sapphire laser, *Opt Lett* **17** 1219-21.
[14] Emmerichs U, Bakker H and Kurz H 1994 Generation of high-repetition rate femtosecond pulses tunable in the mid-infrared, *Opt Comm* **111** 497-501.
[15] Radzewicz C, Pearson G W, Krasinski J S 1993 Use of ZnS as an additional highly nonlinear intracavity self-focusing element in a Ti: sapphire selfmodelocked laser, *Opt Comm* **102** 464-8.
[16] Asaki M T and, Huang C P 1993 Generation of 11-fs pulses from a self-mode-locked Ti:sapphire laser, *Opt Lett* **18** 977-9.
[17] Kafka J, Watts M and Pieterse J 1992 Picosecond and femtosecond pulse generation in a regeneratively mode-locked Ti: sapphire laser, *Optical and Quantum Electronics IEEE* **28** 2151-62.
[18] Fork R L, Martinez E and Gordon J P 1984 Negative dispersion using pairs of prisms, *Opt. Lett.* **9** 150-2.
[19] Wang X and Hu G 1994 Numerical analysis of beam parameters and stability regions in a folded or ring cavity, *J. Opt. Soc. Am. A* **11** 8.
[20] Kogelnik H and Li T 1996 *Appl. Opt.* **13** 1550.
[21] Sherri R E 1998 Analytic expressions for group-delay dispersion and cubic dispersion in arbitrary prism sequences, *J. Opt. Soc. Am. B* **15** 1224-30.
[22] Baatarchuluun Ts and Yang B K, (2007) A sub-10-fs Ti: sapphire oscillator with a simple four-mirror cavity, *J. Korean Phys. Soc* **52** (4) 1043-104.
[23] Haus H A, 2000 Mode-locking of lasers, *IEEE J. Sel. Topics Quantum Electron.* **6** 1173-85.
[24] Meier B and Penzkofer A 1989 *Appl. Phys. B* **49** 513.

IOP Conf. Series: Materials Science and Engineering **704** (2019) 012009 doi:10.1088/1757-899X/704/1/012009

Nanostructured composite materials based on titanium dioxide doped by Mn: template zol-gel synthesis, structure, properties

V V Zheleznov[1,*], A Yu Ustinov[1,2], I A Tkachenko[1], A A Kvach[1], T A Kaidalova[1] and A B Podgorbunsky[1]

[1]Institute of Chemistry FEB RAS, 159 Pr. 100-letiya Vladivostoka, Vladivostok 690022, Russia
[2]Far Eastern Federal University, 8 Sukhanova St., Vladivostok 690091, Russia

*E-mail: zheleznov_sergey@mail.ru

Abstract. Nanostructured composites based on TiO_2 doped with Mn were obtained by sol-gel template method. The morphology and structure of the composites were studied by x-ray diffraction, scanning and transmission microscopy, X-ray photoelectron spectroscopy. Electrical conductivity and magnetic properties are investigated. It is shown that the doping of TiO_2 samples with manganese improves conductivity by two orders of magnitude in comparison with a non-doped sample. Changing the annealing conditions of the nanostructured composite increases the conductive properties four times. It is established that the studied samples have ferromagnetic ordering at room temperature, and the saturation magnetization value and the slope angle of the curve M(H) depend on the annealing conditions, the presence of a magnetic transition in the region of 45 K corresponding to the transition Mn_3O_4 to the ferrimagnetic state is determined.

1. Introduction

In the last decade, a huge number of works on nanoscale titanium oxides have been published. At least in 2014, Chemical Reviews issued 21 reviews on the production, properties, and applications of nanoscale titanium oxides [1, 2]. Investigation of magnetic properties began intensively after the possibility of obtaining a magnetic TiO_2 based semiconductor with room ferromagnetism was shown in [3]. In this paper we investigated the effect of annealing conditions on composition, structure, magnetic and conductive properties of Mn doped nanostructured titanium dioxide microtubes obtained by sol-gel method.

2. Synthesis of materials and methods of study

Nanostructured TiO_2 microtubes doped with manganese were obtained according to the procedure described in [4]. The samples were annealed in air at a temperature of 500° C for 2 hours on a titanium substrate (Sample N1). After annealing the carbon template, the sample was additionally annealed for 2 hours in a porcelain crucible in air at 500° C (Sample N2). The Mn content in the samples (in terms of MnO_2) was ≈17 wt. %).

The surface morphology was studied on a Hitachi S5500 scanning electron microscope (SEM) (Japan) and a Libra 200FE transmission electron microscope (TEM) (Carl Zeiss, Germany). X-ray phase analysis (XPA) was carried out on a D8 ADVANCE diffractometer (Germany) using the PDF-2

IOP Conf. Series: Materials Science and Engineering **704** (2019) 012009 doi:10.1088/1757-899X/704/1/012009

powder database. The magnetic characteristics of the specimens were measured on an MPMS-XL-5 SQUID magnetometer (Quantum Design, United States). X-ray photoelectron spectroscopy (XPS) was carried out using a surface analysis complex of SPECS Company (Bruker, Germany). Electrophysical characteristics of the materials under study were determined by the impedance spectroscopy method (IS) using a Solartron 1260 FRA and DI 1296.

3. Results and Discussion

According to the data of scanning and transmission (Fig. 1a, b) microscopy, the structure of the composite represents micro-sized tubes consisting of nanoparticles. According to the results of X-ray phase analysis, the samples contain anatase, rutile, and also Mn_3O_4 and $MnTiO_3$ phases.

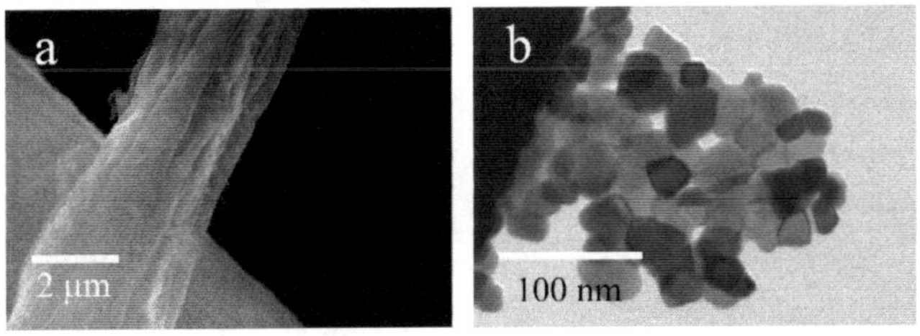

Figure 1. SEM-images of Sample N1.

It should be noted that a change in the annealing conditions leads to a change in the ratio of the phases of rutile-anatase in the composite. As follows from the data of X-ray photoelectron spectroscopy (table 1) oxygen has at least two modifications - the dominant metal oxide and "oxycarboxylic". The content of the metal oxide form is related to the amount of metals and its oxidation states. The presence of carbon is determined by the conditions of template synthesis. Titanium is present in the surface layer in two states: Ti^{+4} and Ti^{+3}.

Table 1. Binding energies (eV) of elements in samples

	Mn ($2p_{3/2}$)	O (1s)		Ti ($2p_{3/2}$)		C (1s)		
MnOT-2	641.5	531.7	529.7	458.5	456.9	288.5	286.3	285.0
MnOT-4	641.4	531.1	529.2	458.4	456.9	288.6	286.3	285.0
	Mn^{3+}	CO_X	TiO_X	Ti^{4+}	Ti^{3+}	CO_2	CO	CH

According to impedance spectroscopy data, the electrical conductivity of undoped TiO_2 reached $0.8 \cdot 10^{-12}$ S/cm. Doping of TiO_2 samples with manganese improves the conductivity by two orders of magnitude compared to the undoped sample (up to $3.8 \cdot 10^{-10}$ S/cm) - Fig. 2.

One of the reasons for the increase in conductivity may be the presence of the pyrophanite $MnTiO_3$ phase, which has highly conductive properties compared to the Mn_3O_4 and TiO_2 phases. However, it should not to exclude the possibility of the formation of a substitution solution when doping titanium dioxide with manganese ($Ti^{4+} \rightarrow Mn^{4+}$). In this case, if Ti^{4+} is replaced by Mn^{4+}, oxygen vacancies should be formed, which can also lead to an increase in electrical conductivity. Thus, the improvement in electrical conductivity can be associated both with the generation of free charge carriers and with the formation of a highly conductive $MnTiO_3$ phase.

IOP Conf. Series: Materials Science and Engineering **704** (2019) 012009 doi:10.1088/1757-899X/704/1/012009

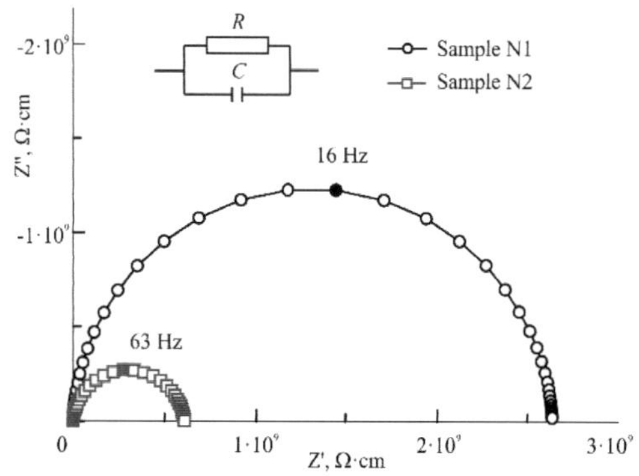

Figure 2. Impedance spectra of Sample N1.

Changing the conditions of annealing of the nanostructured composite allows us to further increase the conductive properties by more than four times ($\sigma = 3.8 \cdot 10^{-10} \rightarrow 1.6 \cdot 10^{-9}$ См/см) (Fig. 2). The results of the investigation of magnetic properties are shown in figures 3, 4 and 5. A characteristic feature of the magnetization field dependences obtained at room temperature is the presence of hysteresis, indicating a ferromagnetic ordering in the samples (Fig. 3). Moreover, an annealing temperature increasing leads to magnetization values growth and changing the curve angle.

The reason for the ferromagnetic ordering at room temperature in paramagnetic titanium dioxide, doped with 3d metal researchers explain using a widespread model of coupled polarons [5]. However, it was shown in [6] that the presence of ordered defects of various types in the TiO_2 structure can also have a strong effect on magnetic properties. The presence of doping manganese ions and defective structure due to annealing makes it possible, in our case, the implementation of the two mechanisms described above for the initiation of ferromagnetic ordering. An increase in the slope of the magnetization curve with a change in the annealing conditions, obviously due to an increase in number of vacancies [7].

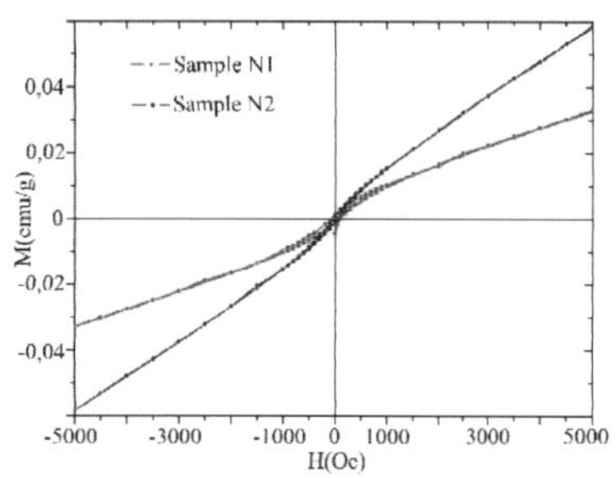

Figure 3. Field dependences of magnetization at 300 K.

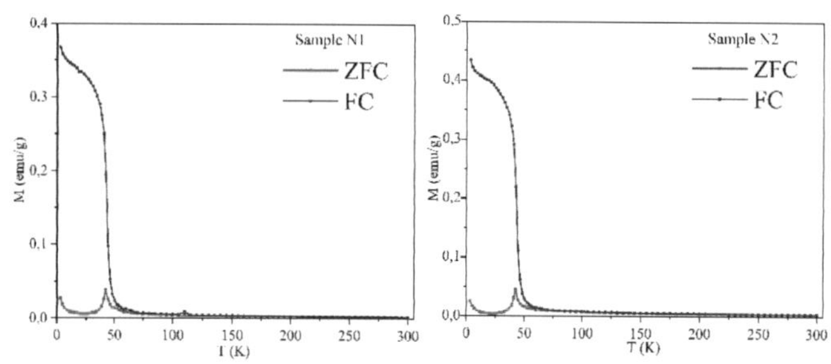

Figure 4. Temperature dependences of magnetization.

The observed peak on the ZFC curve is about 43 K and a sharp increase in the magnetization values in this temperature range on the FC curve (Fig. 4) allows us to say that the samples contain manganese oxide Mn_3O_4. This is because the transition temperature to the ferromagnetic state for Mn_3O_4 coincides with the values we obtained.

The observed shift of the hysteresis loop on the M(H) curves obtained at 3 K (Fig. 5) toward negative fields, indicates that in the studied samples there is a ferromagnet / antiferromagnet type interaction.

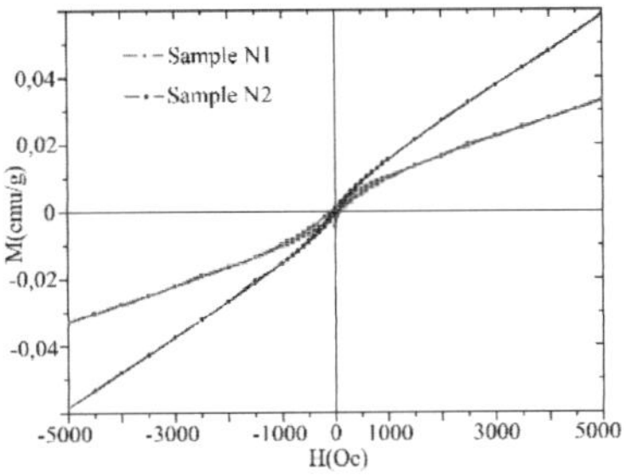

Figure 5. Field dependences of magnetization at 3 K.

It was noted in [6], that defective regions with different spin orders could be formed on the surface of TiO_2 nanoparticles, leading to strong antiferromagnet / ferromagnet type exchange interactions, which most probably take place in our case.

4. Conclusions

Nanostructured titanium oxides doped with manganese have been obtained. The structure of composites was studied. It is shown that doping and changing the conditions of annealing of a nanostructured composite can increase the conductive properties by several orders of magnitude. It was established that the studied samples have ferromagnetic ordering at room temperature. Moreover, the saturation magnetization value and the slope of the M(H) curve depend on the annealing conditions. A relation between the increase in magnetization below 45 K and the transition to the ferromagnetic state of Mn_3O_4 was established. Variation in electrical conductivity values and ferromagnetic ordering at room temperature can be due to the presence of defects in the surface layers with a high probability.

5. Acknowledgements

The synthesis and study of the materials' morphology and structure were carried out within the scopes of the State Order of the Institute of Chemistry FEB RAS (project No. 0265-2019-0001). The electrophysical studies were carried out with the support of the State Order of the Institute of Chemistry FEB RAS (project No. 0205-2019-0005).

6. References

[1] Asahi R, Morikawa T, Irie H and Ohwaki T 2014 *Chemical Reviews* **114** 9824-9852,

[2] Dahl M, Liu Y and Yin Y 2014 *Chemical Reviews* **114** 9853-89,

[3] Matsumoto Y, Murakami M, Shono T, Hasegawa T, Fukumura T, Kawasaki M, Ahmet P, Chikyow T, Koshihara S and Koinuma H 2001 *Science* **291** 854-6,

[4] Zheleznov V.V, Sushkov Y V , Voit E I , Sarin S A and Dmitrieva E É 2015 *Journal of Applied Spectroscopy* **81**(6) 983-9,

[5] Coey J M D, Venkatesan M and Fitzgerald C B 2005 *Nature Materials* **4** 173-9,

[6] Ermakov A E, Uimin M A, Korolev A V, Volegov A S, Byzov I V, Shchegoleva N N and Minin A S 2017. *Physics of the Solid State* **59** (3) 469-82,

[7] Brookes I M, Muryn C A and Thornton G 2001 *Physical Review Letters* **87** 266103.

IOP Conf. Series: Materials Science and Engineering **704** (2019) 012010 doi:10.1088/1757-899X/704/1/012010

Periodic temperature field in a composite plate

P L Abiduev [1], T G Darmaev [2], A S Buldaev [2]
[1]Department of natural-science disciplines, Buryat State Academy of Agriculture, 8 Pushkin street, Ulan-Ude, Russia
[2]Institute of Mathematics and Informatics, Banzarov Buryat State University, 24a Smolina Street, Ulan-Ude, Russia

E-mail: apl087@yandex.ru

Abstract. The temperature fields in two-layer plates under periodic temperature effects on one of the surfaces are considered. The transition to a steady-state temperature field is investigated. For numerical calculations, a program based on the implementation of the finite element method was used. The same steady-state periodic temperature field was obtained asymptotically and was compared with the results of a numerical experiment. Numerical tests showed that for loads close to explosion during detonation of gas mixtures in thermodynamic treatment measures, the periodic temperature field is established after 5-6 cycles of heat flow exposure. Also after analysis of temperature field in single-layer and double-layer plates it is shown that introduction of thermal protection allows to significantly lower temperature of surface under influence of heat flow.

1. Introduction

On the inner surface of the two-layer plate, a periodic heat flux with period T is set, while the other surface is maintained at a constant temperature. Out of physical considerations, it is obvious that after a certain number of cycles of exposure to the heat flux, the temperature field on the inner surface reaches a steady temperature. Studies of the transition process to a steady-state temperature regime were carried out based on a numerical experiment. From a practical point of view, it is of interest to estimate the time it takes to establish a periodic temperature field.

Further, the same steady-state periodic temperature field was obtained asymptotically. The obtained steady-state periodic temperature field is compared with the results of a numerical experiment.

2. Problem statement

The problem statement is as follows:

$$\begin{cases} \dfrac{\partial t_1}{\partial \tau} = a_1 \dfrac{\partial^2 t_1}{\partial x^2}, & 0 \le x \le h, \\ \dfrac{\partial t_2}{\partial \tau} = a_2 \dfrac{\partial^2 t_2}{\partial x^2}, & h \le x \le H, \end{cases} \tag{1}$$

under initial conditions:

$$\begin{cases} t_1(x,0)=0, & 0\le x\le h, \\ t_2(x,0)=0, & h\le x\le H, \end{cases} \tag{2}$$

and conditions on inner surfaces:

$$\begin{cases} \lambda_1 \dfrac{\partial t_1(0,\tau)}{\partial x} = -F(\tau), \\ t_2(H,\tau)=0, \end{cases} \tag{3}$$

when $F(\tau+nT)=Q(\tau)$, $n=0,1,2...$ - periodical heat flux.

At the interface $(x=h)$ it is assumed that ideal thermal contact conditions are set:

$$\begin{cases} \lambda_1 \dfrac{\partial t_1(h+0,\tau)}{\partial x} = \lambda_2 \dfrac{\partial t_2(h-0,\tau)}{\partial x}, \\ t_1(h+0,\tau)=t_2(h-0,\tau). \end{cases} \tag{4}$$

Notations are introduced: $t_i(x,\tau), i=1,2$, - temperature field in the first and second layers respectively, τ - time, x - spatial coordinate. h - inner layer thickness, H - plate thickness, $a_i, i=1,2$ - thermal diffusivity, $\lambda_i, i=1,2$ - coefficients of thermal conductivity of the layers.

3. Asymptotic solution

We obtain the steady-state periodic temperature field in an asymptotic way, without considering the transient process. We use a method based on the application of generalized functions [1,2].

Equations (1) are written:

$$\frac{\partial^2 t}{\partial x^2} = \frac{\partial t}{\partial \tau}\left[\frac{1}{a_1} + \left(\frac{1}{a_2} - \frac{1}{a_1}\right) S_-(x-h)\right] + \left(1 - K_\lambda \frac{\partial t}{\partial x}\right)\bigg|_{x=h} \cdot \delta_-(x-h) \tag{5}$$

Here, $t(x,\tau) = \begin{cases} t_2(x,\tau), \ x \ge h \\ t_1(x,\tau), \ x < h \end{cases}$, $K_\lambda = \dfrac{\lambda_2}{\lambda_1}$, $S_-(x) = \begin{cases} 1, \ x \ge 0 \\ 0, \ x < 0 \end{cases}$ - asymmetric unit function, $\delta_-(x)$ - Dirac delta function. On external surfaces, the conditions (3) are set.

In the asymptotic approach, we assume that the process is sufficiently far from the initial one and that the initial conditions do not influence the temperature distribution. We expand the periodic heat flux in the Fourier series by $[0,T]$:

$$F(\tau) = Q_0 + \sum_{n=1}^{\infty}\left(a_n \cos w_n\tau + b_n \sin w_n\tau\right) = Q_0 + \sum_{n=1}^{\infty}\left(A_n \cos(w_n\tau - \xi_n)\right) \tag{6}$$

Here, $A_n = \sqrt{a_n^2 + b_n^2}$, $\xi_n = \arccos\left(\dfrac{a_n}{b_n}\right)$, $Q_0 = \dfrac{1}{T}\int_0^T Q(\tau)d\tau$, $w_n = n\dfrac{2\pi}{T}$,

$$a_n = \frac{2}{T}\int_0^T Q(\tau)\cos w_n\tau d\tau, \ b_n = \frac{2}{T}\int_0^T Q(\tau)\sin w_n\tau d\tau, \quad n=1,2,3....$$

The solution to problem (5), (3) can be obtained as a superposition of solutions corresponding to the terms in (6), i.e.:

$$t(x,\tau) = t_0(x) + \sum_{n=1}^{\infty} t_n(x,\tau) \tag{7}$$

Here $t_0(x)$ is the static component of the steady-state temperature field, it is defined as the solution of the static heat conduction problem with boundary conditions (3) and the condition on the interface (4) and has the form:

$$t_0(x)=\begin{cases}\dfrac{Q_0}{\lambda_1}(x-h)+\dfrac{Q_0}{\lambda_2}(h-H), & 0\le x\le h,\\ \dfrac{Q_0}{\lambda_2}(x-H), & h\le x\le H.\end{cases} \quad (8)$$

Temperature waves $t_n(x,\tau)$ satisfy the heat equation (5) and the next boundary conditions:

$$\begin{cases}\lambda_1\dfrac{\partial t_n(0,\tau)}{\partial x}=-A_n\cos(w_n\tau-\xi_n)\\ t_n(H,\tau)=0\end{cases}$$

The solution to this problem is found as the real part of the expression:

$$v_n(x,\tau)=X_n e^{i(w_n\tau-\xi_n)} \quad (9)$$

Substituting (9) into (5), we obtain the following problem for determining $X_n(x)$:

$$\frac{\partial^2 X_n}{\partial x^2}=-iw_n\left[\frac{1}{a_1}+\left(\frac{1}{a_2}-\frac{1}{a_1}\right)S_-(x-h)\right]X_n=\left(1-K_\lambda\frac{\partial X_n}{\partial x}\right)\Bigg|_{x=h}\cdot\delta_-(x-h) \quad (10)$$

with boundary conditions:

$$X_n(0)=A_n,\quad X_n(H)=0. \quad (11)$$

With this representation, the obtained solution for the parabolic equation has the character of a boundary layer; therefore, the second boundary condition (11) can be replaced by a simpler one: $X_n(H)\to 0$ while $x\to\infty$.

Solving equation (10) under boundary conditions (11):

$$X_n=\frac{A_n}{\Delta(0)}\left[\Delta(x)S_+(h-x)+q_{1n}e^{-q_{2n}(x-h)}S_-(x-h)\right], \quad (12)$$

where $\Delta(x)=q_{1n}ch(q_{1n}(x-h))-K_\lambda q_{2n}sh(q_{2n}(x-h))$, $q_{jn}=k_{jn}(1+i)$, $k_{jn}=\sqrt{\dfrac{w_n}{2a_j}}$, $j=1,2$,

$S_+(h-x)=1-S_-(x-h)$, i - imaginary unit.

Substituting (12) into (9) and isolating the material part, we determine the temperature waves. In the first layer:

$$t_{1n}(x,\tau)=\frac{A_n}{\sqrt{2}k_{1n}D}\{(k_{1n}^2-k_\lambda^2)[e^{k_{1n}(x)}(\cos(y_{1n}(x)+z_n(\tau))-e^{-k_{1n}(x)}(\cos(y_{1n}(x)-z_n(\tau))]+$$
$$+(k_{1n}-k_\lambda k_{2n})^2[e^{-y_{1n}(x)}(\cos(k_{1n}(x)-z_n(\tau))-e^{y_{1n}(x)}(\cos(k_{1n}(x)+z_n(\tau))]\} \quad (13)$$

In the second layer:

$$t_{1n}(x,\tau)=\frac{A_n\sqrt{2}}{D}\{(k_{1n}+k_\lambda k_{2n})e^{-y_{2n}(x)+k_{1n}h}\cos(-y_{2n}(x)+k_{1n}h+z_n(\tau))-$$
$$-(k_{1n}-k_\lambda k_{2n})e^{-y_{2n}(x)-k_{1n}h}\cos(-y_{2n}(x)+k_{1n}h+z_n(\tau))\} \quad (14)$$

where $D=e^{2k_{1n}h}(k_{1n}+k_\lambda k_{2n})^2+e^{-2k_{1n}h}(k_{1n}-k_\lambda k_{2n})^2+2(k_{2n}^2k_\lambda^2-k_{1n}^2)\cos(2k_{1n}h)$,

$y_{1n}(x)=k_{1n}(x-2h)$, $y_{2n}(x)=k_{2n}(x-h)$, $z_n(\tau)=w_n\tau-\xi_n-\dfrac{\pi}{4}$.

4. Calculations

For numerical calculations of problem (1)-(4) a program based on the implementation of the finite element method was used. The plate is approximated by a grid consisting of one layer of 100 flat isoparametric 4-nodal finite elements. The first layer ($h=H/10$) accounts for 60 elements. This grid turned out to be the most optimal for the convergence of the numerical solution (the numerical solution on a coarser grid (the number of elements is less than 100) has a significant discrepancy with the exact solution). The optimal time step, also selected to take into account the rapidly changing temperature process, was $\Delta\tau = 5\cdot10^{-3}$ s. Numerical calculations were carried out up to the point in time necessary to establish a periodic temperature field. They showed that this temperature field is set for given loads almost after 5-6 cycles of exposure to heat flux.

The calculation results are presented on figure 1 in the form of time dependences on the temperature of the inner surface of a single-layer (dashed line 1) and double-layer plate (solid line 2). Peaks of the curves for the considered time interval (100s equal 5 cycles) go beyond certain stationary level. The maximum values for a single-layer plate and double-layer plate are $430\,^0$C and -$220\,^0$C respectively.

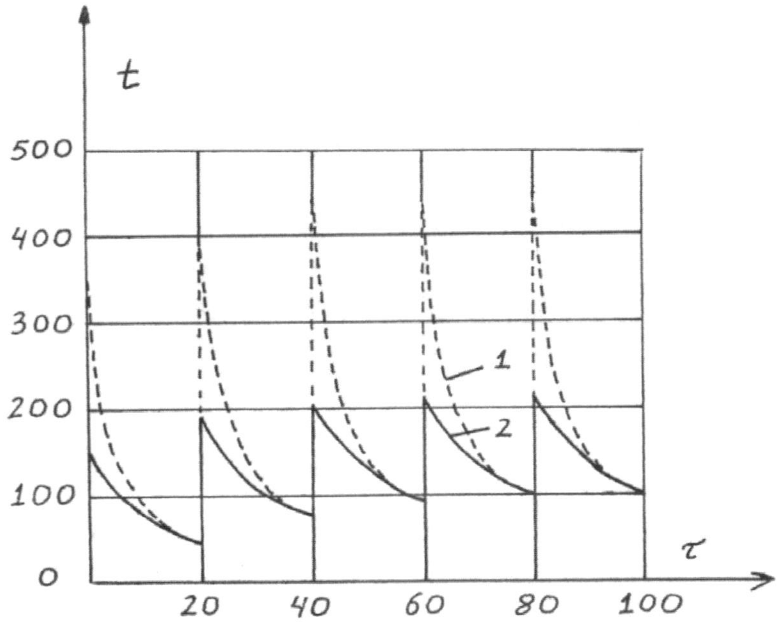

Figure 1.

The expression for the heat flux was taken in the form [3]:

$$Q(\tau) = Q_1(1-\theta)e^{-k_1\sqrt{\tau}} + Q_2 e^{-k_2\tau}.$$

The frequency of the heat flux $T = 20s$.

Constants used in the calculations:

$$Q_1 = Q_2 = 10^3\frac{w}{cm^2},\ k_1 = 4.45\,\mathrm{s}^{-1},\ k_2 = 132.5\,\mathrm{s}^{-1},\ a_1 = 1.16\frac{cm^2}{\mathrm{s}},\ a_2 = 0.16\frac{cm^2}{\mathrm{s}},$$

$$\lambda_1 = 0.96\frac{cal}{cm\cdot \mathrm{s}\cdot {}^\circ\mathrm{C}},\ \lambda_2 = 0.13\frac{cal}{cm\cdot \mathrm{s}\cdot {}^\circ\mathrm{C}},\ \theta = 0.12,\ h = 0.4cm,\ H = 4cm.$$

Numerical calculations using asymptotic formulas (13) - (14) are presented in figure 2 in the form of time dependences on the temperature of the inner surface of a single-layer (dashed line 1) and double-layer (dashed line 2) plates.

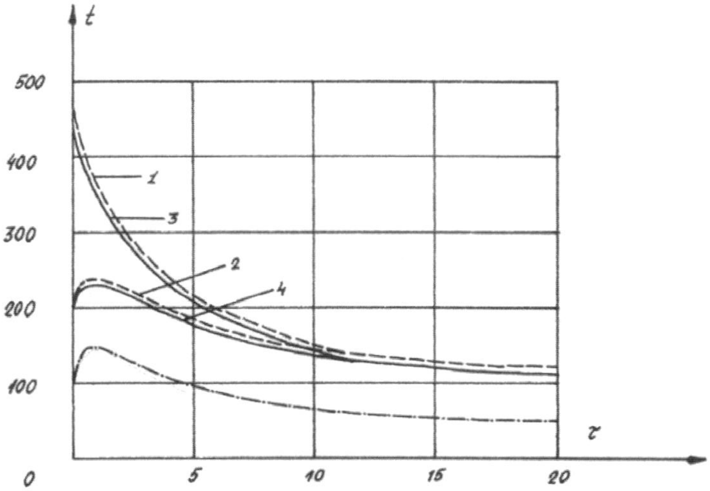

Figure 2.

The results of a numerical experiment are also presented here: solid lines 3 and 4 for a single-layer and double-layer plate, respectively. The considered time interval amounted to $[0,T]$. The dashed-dotted line corresponds to the change in temperature of the inner surface of the double-layer plate upon the first exposure to heat flux.

A slight overestimation of the temperature values obtained by formulas (13) - (14) can be explained by the fact that the plate wall thickness was assumed to be infinite in determining the temperature waves.

The disadvantage of the formulas (13) - (14) is due to the poor convergence of the Fourier series (6) in the expansion of the heat flux function $F(\tau)$, which can be explained by the fact that this function $[0,T]$ does not experience a first-order discontinuity and is rapidly decreasing.

5. Conclusions

Thus, asymptotic expressions have been obtained for determining steady-state periodic temperature fields in double-layer plates when exposed to periodic thermal flux on one of its surfaces. The transition process has been studied numerically using the finite element method. Numerical tests showed that for loads close to explosion during detonation of gas mixtures in thermodynamic treatment measures [3], the periodic temperature field is established after 5-6 cycles of heat flow exposure. Also after analysis of temperature field in single-layer and double-layer plates it is shown that introduction of thermal protection allows to significantly lower temperature of surface under influence of heat flow.

Acknowledgments

The work was supported by the Ministry of Education and Science of the Russian Federation, the grant 1.5049.2017/BCh and Russian Foundation for Fundamental Research, project 18-41-030005.

References

[1] Podstrigach Y S, Lomakin V A and Kolyano Yu M 1989 Thermoelastic bodies of non-uniform structure (Moscow: Science).
[2] Vladimirov V S 1979 Generalized functions in mathematical physics (Moscow: Science).
[3] Adodin S M and Manjalei V I 1986 *Dynamics of solid medium* **74**.

Magnesium thermal nano-sized powders of the chromium subgroup metals

V N Kolosov, V M Orlov, M N Miroshnichenko and T Yu Prokhorova

Tananaev Institute of Chemistry - Subdivision of the Federal Research Centre Kola Science Centre of the Russian Academy of Sciences, 184209, Apatity, Murmansk region, Russia

E-mail: v.kolosov@ksc.ru

Abstract. A method for magnesium vapor reduction of the double oxides of chromium subgroup metals at low static pressure and dynamical vacuum was used. The study results of the production and investigation of tungsten, molybdenum, chromium, W-Cr mix and Mo-W alloy powders are presented. The specific surface area of the resulting powders was in the range from 14 to 41 $m^2 \cdot g^{-1}$. The average crystalline grain size estimated from the value of the range of coherent X-ray scattering was 8-40 nm.

1. Introduction

One of the methods for producing refractory metals is metallothermal reduction of oxide compounds. Due to the high melting point, the product of the reduction of oxide compounds is usually a metal powder. A significant increase in the specific surface area of the powder can be achieved by the reduction of double metal oxides with magnesium vapor [1, 2]. This is due to the fact that the reduction occurs in the particle volume and the reduction product is an alternation of MgO interlayers and metal particles [3]. The presence of MgO or CaO oxide in the precursor increases the number of interlayers. The number of pores increases and, accordingly, the specific surface of the powders increases. At the same time, when oxide compounds of metals of the chromium subgroup were reduced by magnesium vapor, the separation of reaction products with the formation of dense layers of the MgO reaction product outside the reaction zone was found [4, 5]. Moreover, the mass of MgO, located outside the reaction zone increases with increasing residual pressure in the reactor. Removal of MgO from the particle structure leads to decreasing in the porosity of the metal powder and, accordingly, decreasing in its specific surface area [4]. A high temperature of the process also leads to decreasing in the surface of the powders due to coagulation of particles of the reduced metal.

The aim of this work is to study the possibility of obtaining nanosized powders of metals of the chromium subgroup by reduction of their double oxide compounds with magnesium vapors at reduced temperature and pressure.

2. Experimental

Powders of $MgMoO_4$, $CaMoO_4$, $MgWO_4$, $MgCr_2O_4$, $CaW_{0.35}Cr_{0.65}O_4$, $MgW_{0.3}Mo_{0.7}O_4$ and $CaW_{0.3}Mo_{0.7}O_4$, which were synthesized according to the procedures described in [2, 5, 6], were used as precursors. The process was carried out under an atmosphere of magnesium vapor and an inert gas ("high pure" argon) using an experimental setup schematized in figure 1. A vessel (1) containing magnesium (M95) was mounted on the bottom of a reaction beaker (4). Weighed amounts (5 g) of precursors were loaded into metallic containers (3), which were placed over the magnesium-containing vessel. A shield (5) was placed over the containers. The separation between the surface of the reactants and the shield was 27–30 mm. The reaction beaker was covered with a lid (7), which had an orifice in its center for the thermocouple sheath (10). The beaker was mounted in a stainless steel retort, which was hermetically sealed, pumped down to a pressure of 7–10 Pa, filled with argon, heated to a required temperature, and then again pumped down to a residual pressure not exceeding 5 kPa, or the process was carried out under continuous evacuation of the reactor (residual pressure 150-250 Pa). The reduction process was run in the temperature range 600–750°C for 3–5 h.

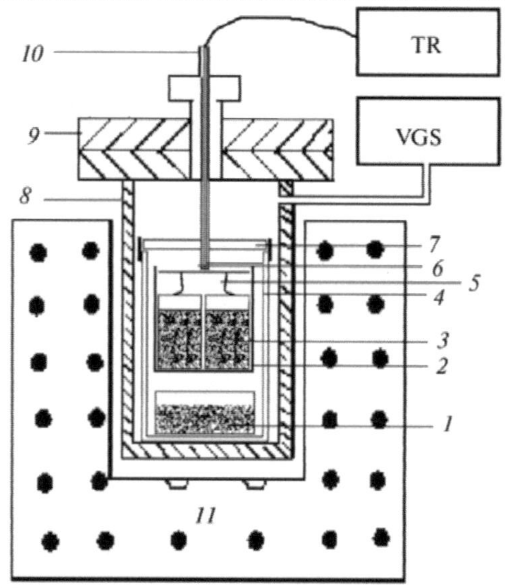

Figure 1. Schematic of the experimental setup: (1) vessel containing magnesium, (2) container for crucibles, (3) crucibles containing the precursor, (4) reaction beaker, (5) shield, (6) thermocouple, (7) lid of the reaction beaker, (8) retort, (9) lid of the retort, (10) thermocouple sheath, (11) electric furnace; TR=temperature recorder, VGS=vacuum–gas system.

The reduction products were treated with a hydrochloric or nitric acid (reagent grade) solution to remove the MgO and CaO. The powder was washed with distilled water until neutral pH was reached and then dried in air at a temperature of 80°C. The phase composition of the powders was determined by X-ray diffraction on a Shimadzu XRD-6000 X-ray diffractometer (CuK_α radiation). The specific surface area and porosity of the powders were determined by BET and BJH measurements on a TriStar II 3020 VI.03 analyzer.

3. Results and discussion

For the precursors listed above, the Gibbs energy ΔG of magnesium reduction reactions was calculated. In the selected temperature range, the calculated ΔG values are in the range -(530–900) $kJ \cdot mol^{-1}$, which confirms the high probability of reduction of the studied binary oxide compounds.

Figure 2 shows crucibles with a reaction product after passing through the corresponding reactions. It is seen that during the process under the dynamic vacuum, the separation of products does not occur. With increasing pressure, the deposition of a white MgO layer begins on the surface of the reaction product and the crucible walls above the loading level. Depending on the composition of the precursor, the separation of the oxide and metal phases begins in the pressure range of 1-5 kPa. Such separation can occur even at temperatures below the melting point of magnesium (figure 1c). The spatial separation of the metal and oxide phases is apparently due to the occurrence of the reduction reaction by the mechanism of the electronically mediated reaction without direct physical contact between the reacting substances [7].

IOP Conf. Series: Materials Science and Engineering **704** (2019) 012011 doi:10.1088/1757-899X/704/1/012011

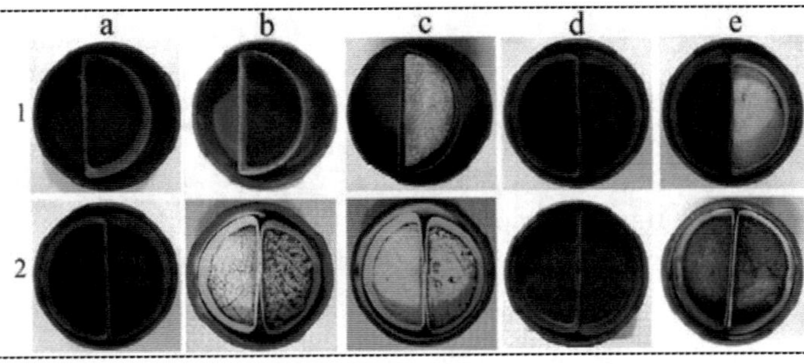

Figure 2. Crucibles with a reaction product after reduction. Precursors: $MgWO_4$ (crucibles 1a, 1b, 1c), $MgW_{0.3}Mo_{0.7}O_4$, (left crucibles 1d, 1e), $CaW_{0.3}Mo_{0.7}O_4$ (right crucibles 1d, 1e), $MgMoO_4$ (left crucibles 2a, 2b, 2c), $CaMoO_4$ (right crucibles 2a, 2b, 2c), $MgCr_2O_4$ (left crucibles 2d, 2e), $CaW_{0.35}Cr_{0.65}O_4$ (right crucibles 2d, 2e); temperature, °C: 1a – 600, 1c, 2a – 630, 1b, 1d, 2b, 2d – 700, 1e, 2c, 2e – 750; residual pressure, kPa: 1a – 0.15, 2a – 0.2, 1b – 0.5, 2b, 2d – 1, 1c, 2c, 2e – 5.

The influence of the reduction conditions and the composition of the precursor on the characteristics of the powders are presented in the table. From the data of the table it can be seen that carrying out the process under these conditions allows completely reducing the selected compounds.

Table 1. The dependence of the characteristics of metal powders on the conditions for the reduction of oxide compounds.

Precursor	Reduction conditions			Product characteristics (PC) and powder properties after leaching (PP)				
	T, °C	P, кПа	t, ч	matter	XRD	FWHM (110), deg.	S_{BET}, $m^2 \cdot g^{-1}$	d_{CSR}, nm
$MgMoO_4$	640	0.5	5	PC	Mo, MgO	0.856	-	-
				PP	Mo	0.812	24.8	24
	720	3	4	PC	Mo, MgO	0.421	-	-
				PP	Mo	0.315	14.7	40
$CaMoO_4$	630	0.15	4	PC	Mo, MgO, CaO	0.892	-	-
				PP	Mo	0.591	25.2	23
	700	1	4	PC	Mo, MgO, CaO	0.491	-	-
				PP	Mo	0.421	17.9	33
$MgWO_4$	640	0.2	5	PC	W, MgO,	2.360	-	-
				PP	W	2.240	41.0	8
	730	5	3	PC	W, MgO,	0.634	-	-
				PP	W	0.523	18.7	16
$MgW_{0,3}Mo_{0,7}O_4$	750	5	5	PC	Mo-W alloy, MgO,	0.693	-	-
				PP	Mo-W alloy	0.439	20.3	19
$CaW_{0,3}Mo_{0,7}O_4$	750	5	5	PC	Mo-W alloy, MgO, CaO	0.436	-	-
				PP	Mo-W alloy	0.331	16.3	27
$MgCr_2O_4$	700	1	4	PC	Cr, MgO	0.472	-	-
				PP	Cr	0.391	24.0	18
$CaW_{0.35}Cr_{0.65}O_4$	750	5	5	PC	Cr, W, MgO	-	-	-
				PP	Cr, W	-	33.0	-

Note: XRD is an X-ray phase analysis, FWHM (110) is the full line width at half maximum, S_{BET} is the specific surface of the powder, d_{CSR} is the average crystallite size calculated from the values of coherent scattering regions.

As a result, powders of tungsten, molybdenum, chromium, a mixture of tungsten with chromium and Mo-W alloy with a specific surface area of 14-41 $m^2 \cdot g^{-1}$ were obtained, which significantly exceeds that one for powders obtained from the same precursors at higher pressures and temperatures [2, 4]. The values of the total width of diffraction lines at half height (FWHM (110)) after washing the powders decrease. Apparently, this is due to the removal of fine metal particles as a result of washing. The adsorption curves of the obtained powders of all metals correspond to type IV according to the IUPAC classification (figure 3).

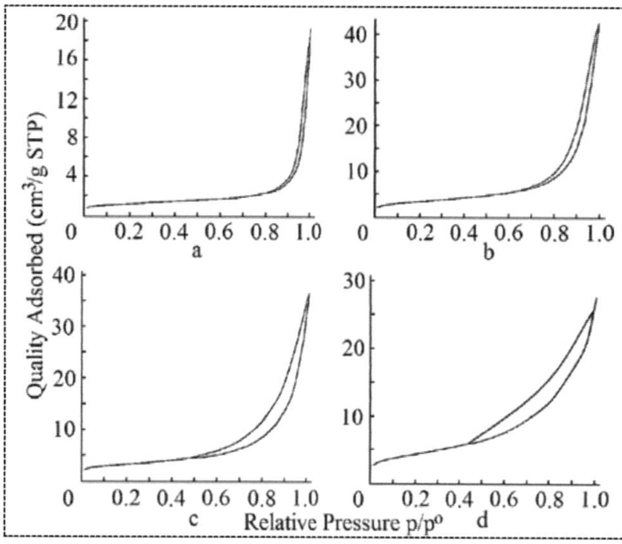

Figure 3. Nitrogen adsorption-desorption isotherms for (a) Mo, (b) Cr (c) W and (d) W-Mo alloy powders. Precursors: (a) $MgMoO_4$, (b) $MgCr_2O_4$, (c) $MgWO_4$, (d) $MgW_{0.3}Mo_{0.7}O_4$; T = 730 °C, t = 4 h.

The specific surface area of the obtained powders is almost completely determined by the pore area, and the specific surface area increases primarily on account of an increase in the amount of small nanopores (figure 4). The larger specific surface area of the powders prepared by reducing $MgW_{0.3}Mo_{0.7}O_4$ in comparison with $CaW_{0.3}Mo_{0.7}O_4$ seems to be due to the removal of a smaller amount of MgO from the reaction zone (figure 2). The removal of a larger amount of MgO from the structure of the reduced particles of the oxide compound leads to a decrease in the porosity of the powder and, accordingly, to a decrease in its specific surface area. This effect is responsible as well for the observed decrease in the specific surface area of the powders with increasing residual pressure in the reactor.

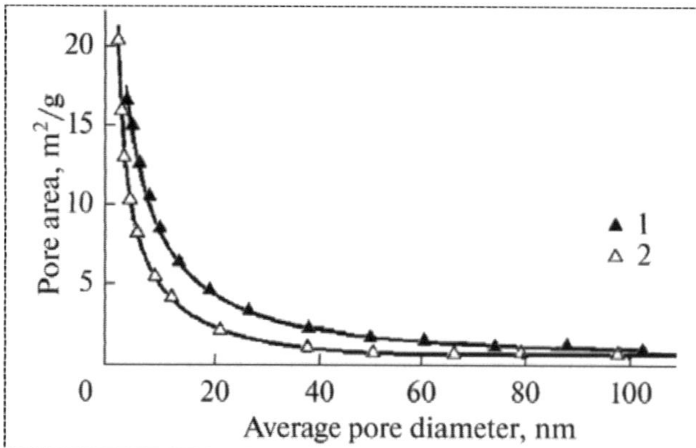

Figure 4. Total pore area as a function of average pore diameter for the Mo-W alloy powders. Precursors: (1) $CaW_{0.3}Mo_{0.7}O_4$, (2) $MgW_{0.3}Mo_{0.7}O_4$; temperature, 750°C; pressure, 5 kPa; time, 5 h; specific surface area of the Mo-W alloy powders: (1) 16.3, (2) 20.3 $m^2 \cdot g^{-1}$.

Thus, the present results demonstrate the feasibility of obtaining nanosized powders of metals of the chromium subgroup with a large specific surface area by reduction of their double oxide compounds with magnesium vapors at reduced temperature and pressure. Note that the specific surface area and nanopore structure of the powders can be controlled by properly selecting the starting oxide compound and reduction conditions.

4. Conclusion

The reduction by magnesium vapor of the oxide compounds of metals of the chromium subgroup under dynamic vacuum and low static rarefaction (0.15-5 kPa) in the temperature range of 600–750 °C was studied.

The lower range of residual argon pressures of 1-5 kPa was established, at which the separation of the metal and oxide phases in the reaction products begins with the formation of dense MgO layers outside the reaction zone. It is apparently due to the reaction proceeding by the electron-mediated reaction mechanism without direct physical contact between reactants.

Powders of W, Mo, Cr, a mixture of W, Cr and a Mo-W alloy with a surface of 15-41 $m^2 \cdot g^{-1}$ were obtained. The crystallite size estimated by the X-ray method is 8-40 nm.

It was determined that the shape of the adsorption curves for the obtained powders can be attributed to type IV according to IUPAC, which is characterized by the presence of a hysteresis loop, and is characteristic of materials with a mesoporous structure.

Acknowledgments

This study was partially supported by the RFBR grant, the project 18-03-00248.

References

[1] Orlov V M, Kryzhanov M V and Kalinnikov V T 2014 Magnesium reduction of tantalum oxide compounds *Dokl. Chem.* **457** 160

[2] Kolosov V N, Orlov V M, Miroshnichenko M N and Prokhorova T Yu 2017 Choice of double oxides for magnesium vapor preparation of nanopowders of chromium subgroup refractory metals *Isvestija S.-PbGTI(TU)* **38** 48

[3] Müller R, Bobeth M, Brumm H, Gille G, Pompe W and Thomas J 2007 Kinetics of nanoscale structure development during Mg-vapour reduction tantalum oxide *Int. J. Mater. Res.* **98** 1138

[4] Orlov V M and Kolosov V N 2016 Magnesiothermic reduction of tungsten and molybdenum oxide compounds *Dokl. Chem.* **468** 162

[5] Kolosov V N, Orlov V M, Miroshnichenko M N and Prokhorova T Yu 2018 Investigation of the obtaining of Mo-W alloy from complex oxides by magnesium vapours reduction *Trans. Kola Science Centre* **9** 285

[6] Miroshnichenko M N, Kolosov V N, Makarova T I and Orlov V M 2017 Synthesis of molybdates and tungstates of calcium and magnesium *Isvestija S.-PbGTI(TU)* **38** 44

[7] Kolosov V N and Orlov V M 2019 Electronically mediated reactions in metal thermal reduction of molybdenum and tungsten oxide compounds *Dokl. Phys. Chem.* **484** 28

IOP Conf. Series: Materials Science and Engineering **704** (2019) 012012 doi:10.1088/1757-899X/704/1/012012

Morphological study of $PM_{2.5}$ by SEM-EDS in Ulaanbaatar

N Tuvjargal[1], L Enkhtsetseg[2], D Shagjjamba[2] and P Zuzaan[2]
[1]Department of Physical Sciences, National University of Mongolia, University street-1, Ulaanbaatar 14201, Mongolia
[2]Nuclear Research Center, National University of Mongolia, Peace avenue-122, Ulaanbaatar, Mongolia

E-mail: tuvjargal@num.edu.mn

Abstract. A morphological and elemental composition of $PM_{2.5}$ in air pollution of Ulaanbaatar by Scanning Electron microscopy (SEM) combined X-ray Energy Dispersive Spectrometry (EDS) were investigated. The morphological parameters were measured using the 5k magnification picture of SEM and analyzed the 752 individual particles of the 25 days (winter-10, summer-15) sample of the $PM_{2.5}$. The size distribution shown the ~45% of particles were most harmful fraction of particulate matters for human health (less 0.4μm). These fraction particles increased up to 92% in winter. It was shown that ultrafine and irregular $PM_{2.5}$ results from the combustion. The results of determining the elemental composition of these individual particles are presented.
Keywords: $PM_{2.5}$, Morphology and elemental composition, Size distribution, SEM, EDS

Introduction

The particulate air pollution is a growing problem in cities with emissions generating activities; these activities are mainly industrial and vehicular widely developed in large cities in different countries. Particulate also called aerosols air pollution is caused due to very small liquid and solid particles suspended in the air. They originate from a variety of stationary and mobile sources and may be directly emitted or formed in the atmosphere by transformation of gaseous emissions.

The use of Scanning Electron Microscope (SEM) has found application in the field of particulate air pollution for several years; its use has helped the study of particle morphology and single particles chemical composition. In fact, the simultaneous characterization of both physical-chemical and morphological parameters of a complex mixture of organic and inorganic particulate matter is one of the major aspects for the characterization and identification of emission sources that contribute to particulate concentrations in the atmosphere [1]. Size and chemical composition of ambient particulates strongly influence on human health, visibility and ecosystem etc.

Thus it is crucial to investigate the physicochemical characteristics of atmospheric particles and also to evaluate their potential toxicity. A large number of health-related studies recognized that fine particles particularly submicron sizes, penetrate deep into lung and exacerbate chronic respiratory and pulmonary diseases [2, 3]. Such diseases also induce morphological and functional alterations in human pulmonary epithelial cells [4]. Apart from this, elemental composition of particulate plays an important role in the chemical characteristics of particulate matters (PM) and provides interesting data, not only for the evaluation of its impact on human health, ecology and environment but also for the identification of specific emission sources [5].

The health effects of particulate emissions depend on the size of the particles. The larger particles emitted by cars do not find their way into the body as easily as the very fine particles emitted by laser printers. The smaller the particles, the more frequently adverse health effect on the entire organ system are observed. Ultrafine particles are classified as highly carcinogenic and are therefore especially harmful. According to studies, particles averaging 10 micrometers or less in size can adversely affect lung function, while particles smaller than 2.5 micrometers in size can trigger systemic diseases such as tumors or cardiovascular problems. The size of the particles is then the characteristic most studied in terms of classification and health damage. However, if other features such as the elemental composition of the particle are considered the risk of health damage would probably be higher [6]. Specially, Particulate matter with aerodynamic diameters less than 2.5μm ($PM_{2.5}$) has been found to cause health problems. It can trigger or exacerbate conditions, such as asthma, emphysema, and bronchitis, silicosis and lung cancer [7-9].

In this context, particles found in environment by the analysis of filters collected in area of the near to Ulaanbaatar city. For Mongolia's air pollution level, annual mean of $PM_{2.5}$ in the capital Ulaanbaatar city are 6-10 times higher than those considered safe by the WHO air quality guidelines [10]. Air pollution has become one of the most challenging issues in Mongolia, exacerbated during winter time because of solid fuel combustion. Especially, the capital city Ulaanbaatar where more than 47% of total population lives in is the most air pollution affected area in the country. In fact, during winter, Ulaanbaatar's air pollution is caused by households and low pressure boilers burning raw coal in Ger district (80%); motor vehicle (10%); coal-fired power plants (6%); and solid waste and soil degradation (4%) [11,12]. In addition to be a large source of ambient air pollution, polluting fuels such as raw coal and biomass use for cooking and heating contribute to a significant additional burden to the health of the population from household air pollution [13].

As the development of tools for more precise characterization of atmospheric aerosols has been steady, as is the case for the use of electron microscope as the impact it has had the use of electrons, the study of particles, has been enormous, since it has allowed to obtain a vast information on the nature, origin and elemental composition of our object of investigation. The scanning electron microscope has sufficient specificity for the analysis of small particles and below the range of nanometers. Scanning microscope combines image of the particle morphology, i.e. shape, size, roughness, etc. as well as analytical information of the chemical elements that comprise the analysis of single atmospheric particles by Scanning Electron Microscopy coupled with Energy Dispersive X-ray (SEM-EDS) is a powerful tool of recognizance emissions in urban places because the SEM may examine particles which are too small to be studied in the conventional optical microscope and also inform on the nature and origin of the particles revealed after the morphology, size and elemental composition [14-19].

The aim of this work is identification and characterization (morphology and elemental composition) of particles present in samples of $PM_{2.5}$ as part of an air quality study by SEM-EDS in Ulaanbaatar.

Sampling and analythical method

Particles found in environment by the analysis of filters collected in atmospheric monitoring station property of the Nuclear Research Center, NUM in the City of Ulaanbaatar. The map shown in Figure 1 shows these locations. The station is located in the southeastern area of the city.

A $PM_{2.5}$ particle was collected by GENT sampler, using a sampling flow of 16liter/min. Sampling was carried during 24 hour for two day in week. The filters used were polycarbonate, for collecting the aerosols samples polycarbonate nuclepore filters 47mm, 0.4μm pore diameters, were used. This type of filters media gives good results for morphological and dimensional analysis, in particular their pure composition and their flat surface let them suitable for the SEM analysis. Analyses of $PM_{2.5}$ were performed using Hitachi SU8010 scanning electron microscopy (SEM) with energy dispersive spectrometer (EDS) in Inner Mongolia Normal University of China.

Specimens were processed by separating the collected particles in the polycarbonate filter which was placed over a sample holder, and is introduced into the chamber of SEM. The images of the samples were taken at a magnification of 1k, 2k, 5k, 10k, 20k, 40k, 45k and 60k. Then 5k magnification SEM images allow to analyses the elemental composition and morphological parameters of particles in the entire particle size range considered. Spectra of individual particles were obtained after scanning an electron beam with an accelerating voltage of 15 kV for determination of individual elemental chemical composition of the particles. Will be identified with EDS spectra the main chemical composition that make up the particles in the study area.

Those images are then elaborated with image analysis software Image-J®, for the dimensional of morphological analysis. Several morphological parameters were determined for $PM_{2.5}$ classification. Count, area, length, breadth and perimeters measurements were automatically obtained using the software. All dimensional parameters were measured in calibrated units (microns).

Fig.1 Study area in UB city, geographic location

Estimations were also made of equivalent spherical diameter (ESD), a commonly used parameter for particle sizing using following equation (Eq. 1);

$$\mathrm{ESD} = 2\sqrt{\frac{Area}{\pi}} \qquad (1)$$

We obtained the sphere and flake shapes and remaining is classified as part of the irregular of these. Irregular shape is no geometrical iconic particles. Thus is obtained three topical shapes.

Results

The average mass concentration of $PM_{2.5}$ samples in summer was found to be ~52.96μg/m^3, in winter was found to be ~101.29μg/m^3. The observed concentration is 2.2 and 4 times higher than the WHO standards (25μg/m^3) and slightly higher than the MNS4585:2016 (50μg/m^3) in summer, 2 times higher in winter. We obtained 480 SEM images of $PM_{2.5}$ particles (15 parts for each filter of 25 days samples) where its size and shape distribution, and elemental concentration of 639 each particles individually was analyzed by SEM-EDS.

Fig.2 shows that particles size distribution on the four size intervals established, indicated by percentages of the total analyzed particles. The size distribution results presented in the 2.95% of the particles identified >1.8μm, 4.85% 1-1.8 μm, 12.65% 0.8-1 μm, 33.9% 0.4-0.8 μm and the remaining 45.3% less than 0.4μm. It is ~45% have been up to 0.4μm which is the most harmful fraction of particulate matters for human health.

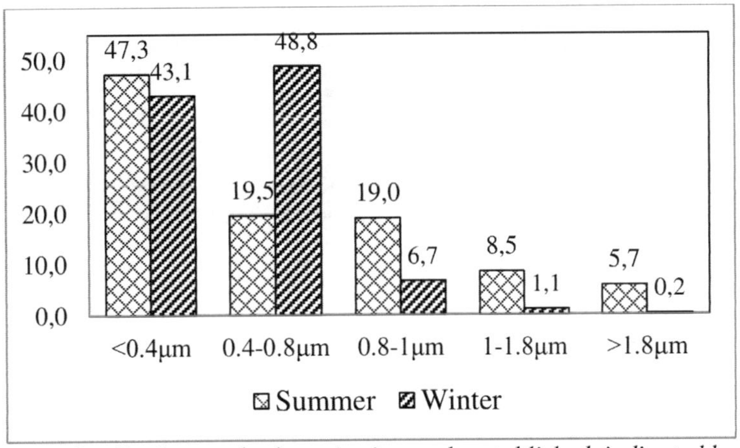

Fig.2 Particles size distribution on the four size intervals established, indicated by percentages of the total analyzed particles, %.

Fig.3 shows that Particles shape distribution on the 3 shape intervals established, indicated by percentages of the total analyzed particles. For the shape distribution, 55% - sphere of the total counted particles, 41% -irregular, and 4% -flake. Irregular shapes dominated in winter and sphere shapes dominated in summer, whereas flake shapes was no seasonal. This results shows that the summer PM pollution is sphere shape from the soil in summer and winter PM pollution is irregular from the combustion, based on the report of the air pollution study of Ulaanbaatar [17]. Irregular shape particles 2 times higher than taking into up to 0.4μm size particles increased to 92% in winter. It is indicates from the combustion that $PM_{2.5}$ is dominated by ultrafine size and irregular shape.

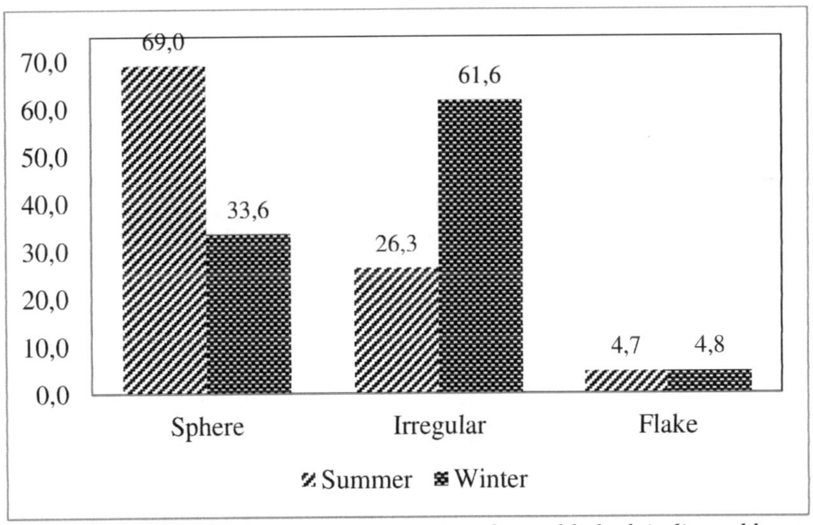

Fig.3 Particles shape distribution on the 3 shape intervals established, indicated by percentages of the total analyzed particles, %.

We obtained 752 particles of $PM_{2.5}$ particles where its chemical composition of each particle individually was analyzed by EDS. Fig.4 indicates compositional data of elemental chemical constituents and counting percentage from $PM_{2.5}$ particles which showed the following elements: Na, Mg, S, Al, Si, Cl, K, Ca, Ti, V, Cr, Mn, Fe, Ni, Cu, Zn, Sr and Pb and the predominant elements were Mg, S, Ca, Fe and Al. Polycarbonate nucleapore filters consist of C and O; therefore those two elements were subtracted from loaded filters. Elemental analysis result shows the summer particles dominated by Na, Mg, S, Al, Cl, K, Ca, Ti and Fe. Whereas winter particles dominated by Si, V, Cr, Mn, Ni, Cu and Pb.

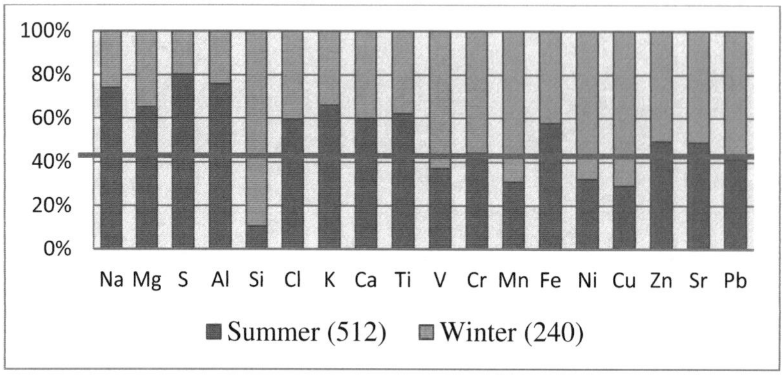

Fig.4 indicates compositional data of elemental constituents and counting percentage in summer 512 and winter 240 individual $PM_{2.5}$ particles.

Conclusion

We investigated a morphological and elemental composition of $PM_{2.5}$ in air pollution of Ulaanbaatar by SEM-EDS. The SEM-EDS technique is a valuable tool for the characterization of PM especially for individual particles. The identification of the morphology and elemental composition of these particles provides valuable information for the determination of their origin and formation processes. The results of the size distribution shown the ~45% have been up to 0.4μm which is the most harmful fraction of particulate matters for human health. For the shape distribution, 55% - sphere of the total counted particles, 41% -irregular, and 4% -flake. Irregular shapes dominated in winter and sphere shapes dominated in summer, whereas flake shapes was no seasonal. This results shows that the summer PM pollution is sphere shape from the soil in summer and winter PM pollution is irregular from the combustion, based on the report of the air pollution study of Ulaanbaatar. These fraction particles increased to 92% in winter which is most of the irregular shape. It has been shown that ultra fine and irregular $PM_{2.5}$ indicates from the combustion. The results of determining the elemental composition of these individual particles are also given.

References

[1] Ramirez-Leal R, Esparza-Ponce H and Duarte-Moller A 2007 Characterization of Inorganic Atmospheric Particles in Air Quality Program with SEM, TEM and XAS. *Revista Mexicana de Fisica*, **53** 102-7.

[2] Pope C A, Burnett R T, Thun M J , Calle E E , Krewski D , Ito K and Thurston G D 2002 Lung Cancer, Cardiopulmonary Mortality, and Long-Term Exposure to Fine Particulate Air Pollution. *The Journal of the American Medical Association* **287** 1132-1141. http://dx.doi.org/10.1001/jama.287.9.1132

[3] Berico M, Luciani A and Formignani M 2007 Atmospheric Aerosol in an Urbana Area Measurement of tsp and PM Standards and Pulmonary Deposition Assessment. *Atmospheric Environment,* **31** 3659-65. http://dx.doi.org/10.1016/S13522310(97)00204-5

[4] Dockery D W and Pope C A (1994) Acute Respiratory Effects of Particulate Air Pollution. *Annual Review of Public Health* **15** 107-32.

[5] http://mccog.net/cleanairdare/air-quality-explained.html

[6] Querol X, Alastuey A, Puicercus J A, Mantilla E, Ruiz C R, Lopez-Soler A, Plana F and Juan R 1998 Seasonal Evolution of Suspended Particles around a Large Coal-Fired Power Station *Chemical Characterization. Atmospheric Environment,* **30**, 3557-72. http://dx.doi.org/10.1016/13522310(96)00108-2

[7] Ramirez-Leal R, Esparza-Ponce H E, Varela-Sortillón A, Astorga-Reyes A and Roman-B A 2009 Elemental Chemical Composition, Size and Morphological Characterization of Individual Atmospheric Particles within an Air Quality Program. *Microscopy and Microanalysis*, 15, (S2) 1300-1301.

[8] Thomas 2006 Assessment of the Health Effects of Toner Particles on People in the Workplace. BG Research Institute for Occupational Medicine. 14

[9] https://www.epa.gov/pm-pollution/particulate-matter-pm-basics

[10] http://www.who.int/phe/publications/air-pollution-global-assessment/en/

[11] Davy P K, Gunchin G, Markwitz A, Trompetter W J, Barry B J, Shagjjamba D and Lodoysamba S 2011 Air particulate matter pollution in Ulaanbaatar, Mongolia: determination of composition, source contributions and source locations. *Atmospheric Pollution Research* 2(2) 126-37.

[12] Guttikunda S K, Lodoysamba S, Bulgansaikhan B and Dashdondog B 2013 Particulate Pollution in Ulaanbaatar, Mongolia *Air Quality, Atmosphere & Health*, **6** (3) 589–601

[13] World Health Organization 2016 Burning opportunity: Clean Household Energy for Health, Sustainable Development, and Wellbeing of Women and Children.

[14] Chung W, Sharifi V N, Swithenbank J, Ossamor O and Nolan A 2008 Characterization of Airborne Particulate Matter in a City Environment. *Journal of Modern Applied Science*, 2, 17-32.

[15] Okada K and Kai K 2004 Atmospheric Mineral Particles Collected at Quira in the Taklamakan Desert, China. *Atmospheric Environment*, **38**, 6927-35. http://dx.doi.org/10.1016/j.atmosenv.2004.03.078

[16] Zhai Y B, Fu Z M and Wang L F 2012 Characteristic, Composition, and Sources of TSP Investigated by HRTEM/EDS and ESEM/EDS. *Environmental Monitoring and Assessment*, **184**, 6693-707. http://dx.doi.org/10.1007/s10661-011-2452-5

[17] Ariola V, D'Alessandro A, Lucarelli F, Marcazzan G, Mazzei F and Nava S 2006 Elemental Characterisation of PM_{10}, $PM_{2.5}$ and PM_1 in the Town of Genova (Italy). *Chemosphere*, **62**, 226-32. http://dx.doi.org/10.1016/j.chemosphere.2005.05.004

[18] Hays D M, Fine M P, Geron D C, Kleeman J M and Gullett K B 2005 Open Burning of Agricultural Biomass:Physical and Chemical Properties of Particle-Phase Emissions. *Atmospheric Environment*, **39**, 6747-64. http://dx.doi.org/10.1016/j.atmosenv.2005.07.072

[19] Semeniuk T A, Wise M E, Martin S.T, Russell L M and Buseck P R 2007 Water Uptake Characteristics of Individual Atmospheric Particles Having Coatings. *Atmospheric Environment*, **41**, 6225-35. http://dx.doi.org/10.1016/j.atmosenv.2007.04.001

IOP Conf. Series: Materials Science and Engineering **704** (2019) 012013 doi:10.1088/1757-899X/704/1/012013

Synthesis of highly dispersed precursor nanopowders for *YAG:Yb* ceramics by the use of hexamethylenetetramine

I S Chikulina[1], A A Kravtcov [1,2], M S Nikova[1], F F Malyavin[1], V A Lapin [1,2], D S Kuleshov[1]

[1]North Caucasus Federal University, 355029, Kulakova st., 2, Stavropol, Russia

[2]Federal Research Center The Southern Scientific Centre of the Russian Academy of Sciences, 344006, Chehova st., 41, Rostov-on-Don, Russia

*iraaaa@yandex.ru

Abstract. The effect of hexamethylenetetramine on the morphology, dispersion composition and chemical homogeneity of the *YAG:Yb* ceramic nanopowders, obtained by the method of heterophase co-precipitation, followed by calcination (1100 °*C*) was studied. It was shown that the use of an aqueous hexamethylenetetramine solution allows obtaining *YAG:Yb* nanopowders with the particles size of 40-70 nm with a degree of agglomeration of 2.54, which is significantly less than that in comparison with the powders, obtained by precipitation in ammonia.

1. Introduction

Ceramics based on yttrium-aluminum garnet, in particular, *YAG:Yb* is one of the promising materials for the optics and solid-state lasers manufacture. Such ceramics should have high transparency and high density. These properties substantially depend on the characteristics of the initial precursor powder: its purity, morphology, and dispersion. The problem of obtaining the monodisperse precursor powder with the required morphology while maintaining the purity and homogeneity of the chemical composition is of great importance in modern materials science in the field of ceramics.

The most commonly used precipitants are an aqueous urea solution, solutions of ammonium bicarbonates and carbonates, and an aqueous ammonia solution [1-5]. Hexamethylenetetramine (hexamine) is a promising precipitant, but the process of precipitation when use hexamine as a precipitant are not well understood. In this regard, the aim of this work was to conduct a comparative analysis of the properties of the precursor powders and *YAG:Yb* ceramic powders obtained via precipitation by aqueous solutions of ammonia and hexamine.

2. Materials and Methods

In this work, for the synthesis of powders of *YAG: Yb* composition, we used a concentrated aqueous ammonia solution and a hexamine solution (1.25 M). The choice of hexamine concentration is based on the results of previous studies [3]. Samples were synthesized by reverse heterophase coprecipitation by spraying a salts solution into an alkaline precipitant solution. The obtained precursors (samples A1 and H1) were dried in air and then ball-milled using the wet method in a Pulverizette 5 planetary mill under the same conditions.

Ceramic powders (samples A2 and H2) were synthesized by high-temperature treatment of A1 and H1 precursor powders in the air at a temperature of 1100 °C in an LHT 08/18 chamber furnace. The morphology of the obtained samples was studied using a MIRA3-LMH scanning electron microscope. The specific surface area was measured using a 3Flex specific surface area and porosity analyzer. To study the particle size distribution, an Analysette 22 MicroTec laser particle size analyzer was used. The phase composition of the samples was determined using an ARL Xtra, TermoScientific X-ray diffractometer. The calculation of the average size of the agglomerates in powders was performed by the following equation:

$$D\_BET=6/(\rho*S\_BET), \quad (1)$$

where ρ – density of YAG: Yb, g/cm^3, SBET – specific surface area, m^2/g.
The size of nanocrystallites was determined using the Scherrer formula:

$$D\_XRD=0.89\lambda/(\beta*\cos\Theta), \quad (2)$$

where λ – CuKα radiation wavelength (0,15406 nm), β – width at half maximum of the diffraction peak.

3. Results and Discussion

The results of X-ray diffraction analysis showed that in both samples, after the precipitation A1 and H1 powders were amorphous, as evidenced by the absence of X-ray reflections in the diffraction patterns. After heat treatment, samples A2 and H2 consisted of a 100% garnet phase.

Studies have shown that the morphology of samples synthesized using ammonia (A1, A2) and hexamine (H1, H2) was significantly different. As follows from the scanning electron microscopy results (Fig. 1), when using ammonia, denser A1 powder was formed, as compared to H1 sample synthesized using hexamine. After high-temperature treatment, fundamental differences in the samples were retained. Ceramic powder A2 had a denser microstructure than H2 sample, which was characterized by much less agglomeration of particles (Fig. 1).

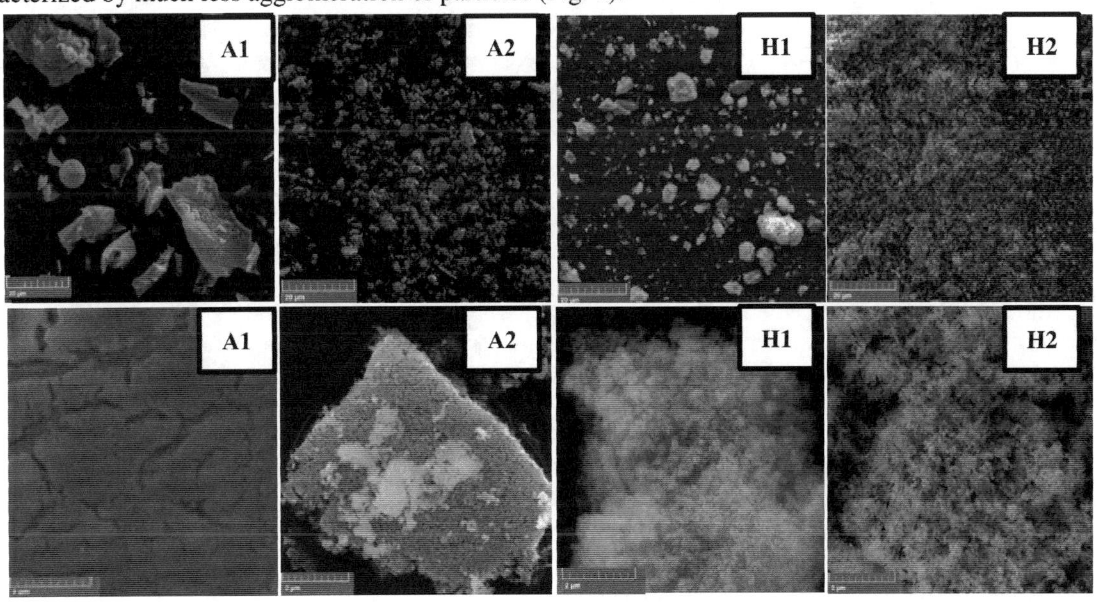

Figure 1. SEM micrographs of the samples:
A1 and H1 – precursor powders; A2 and H2 – ceramic powders

In A2 ceramic powder, obtained from A1 precursor powder "neck bridges" between the particles were formed during the high-temperature processing. As a result, A2 ceramic powder consisted of rigid agglomerates ranging in size from 0.78 to 5.80 μm.

In the case of the use of hexamine as the precipitant, H2 powder had a smoothed surface topography of the particles and a narrower size distribution (Tab. 1). Differences in morphology and particle size distribution reflected in differences in the specific surface area of the samples. Ceramic powder H2 had a higher specific surface area (13.39 m^2 / g) than A2 powder (7.50 m^2 / g).

Analysis of A2 ceramic powder showed that it formed by agglomerates, which size significantly exceeds the average size of crystallites (Tab. 1). Therefore, a high degree of agglomeration (69.17) of these powders is consequential. In H2 sample obtained using hexamine, the number of primary crystallites in the agglomerate is much lower and reaches a value of 2.54.

Tab. 1 – Dispersion characteristics of precursor powders and ceramic powders

Name of sample	Precipitant	Particle size distribution			Specific surface area (S_{BET}), m^2/g	The crystallite size (D_{XRD}), nm	The degree of agglomeration (n)
		D_{10}	D_{50}	D_{90}			
A1	Concentrated ammonia solution	1,27	4,26	13,92	51,57		
H1	Aqueous hexamine solution (1,25 M)	1,09	3,62	11,41	65,72		
A2	Concentrated ammonia solution	0,78	1,95	5,80	7,50	41,00	69,17
H2	Aqueous hexamine solution (1,25 M)	0,17	0,89	2,91	13,39	69,03	2,54

4. Conclusion

According to the results of the study, it was concluded that the use of hexamine allows one to obtain a low agglomerated, highly dispersed powder with a homogeneous chemical and phase composition. This powder characterized by good formability and low-temperature sintering of ceramic compacts, which is an important condition for reducing porosity and improving the optical properties of ceramics. Therefore, it can be argued that hexamine is a promising precipitant for the synthesis of *YAG: Yb* ceramic powders.

References

[1] Malyavin F F, Tarala V A, Kuznetsov S V, Kravtsov A A, Chikulina I S, Shama M S, Medyanik E V, Ziryanov V S, Evtushenko E A, Vakalov D S, Lapin V A, Kuleshov D S, Tarala L V and Mitrofanenko L M 2019 Influence of the ceramic powder morphology and forming conditions on the optical transmittance of YAG:Yb ceramics *Ceram. Int.* **45** (4) 4418-23

[2] Tarala V A, Shama M S, Chikulina I S, Kuznetsov S V, Malyavin F F, Vakalov D S, Kravtsov A A and Pankov M A 2018 Estimation of Sc^{3+} Solubility in Dodecahedral and Octahedral Sites in *YSAG:Yb. J. Am. Ceram. Soc.* **102** (8) https://doi.org/10.1111/jace.16294

[3] Kravtsov A A, Chikulina I S, Tarala V A, Evtushenko E A, Shama M S, Tarala L V, Malyavin F F, Vakalov D S, Lapin V A and Kuleshov D S 2019 Novel synthesis of low-agglomerated YAG:Yb ceramic nanopowders by two-stage precipitation with the use of hexamine. *Ceram.Int.* **45**(1) 1273–82.

[4] Li J., Liu Z, Lei Wu L, Han X, Qi X and Sun X 2018 Influence of ammonium sulfate on *YAG* nanopowders and Yb:YAG ceramics synthesized by a novel homogeneous co-precipitation method. *Journal of Rare Earths* **36** (9) 981-5

[5] Li J-G, Ikegami T, Lee J-H, Mori T and Yajima Y 2000 Co-precipitation synthesis and sintering of yttrium aluminum garnet (*YAG*) powders: the effect of precipitant. *Journal of the European Ceramic Society* **20** 2395-2405.

IOP Conf. Series: Materials Science and Engineering **704** (2019) 012014 doi:10.1088/1757-899X/704/1/012014

Acoustic Method of Division of Ultradispersible Powders

T G Darmaev [1], S V Kalashnikov [2], A V Nomoev [3]

[1]Institute of Mathematics and Informatics, Banzarov Buryat State University, 24a Smolina Street, Ulan-Ude, Russia
[2] Center of Shared Facilities, Banzarov Buryat State University, 24a Smolina Street, Ulan-Ude, Russia
[3]Institute of Physical Material Science of Siberian Branch of Russian Academy of Science, 6 Sakhyanov Street, Ulan-Ude, Russia

dtg@bsu.ru

Abstract. It is known that nanodispersible powders are the condensed medium which at the same time have properties of liquid and solid bodies. And properties of such an environment strongly depend on particle size. In this work, the acoustic method of division of ultradispersible powders on the fluctuating plate is considered. The experimental data by the efficiency of the division of nanoparticles of dioxide of silicon are obtained. Based on a method of successive approximations of the Navier-Stokes equations for vorticity the equations and boundary conditions for the fluctuating plates taking into account amplitude to the second-order are received. Using the Mathematica system, analytical decisions are received and plots of vectors fields of speeds for a plate of the infinite length are constructed.

1. Introduction

The majority of methods of production of nanopowders give a product with very wide distribution by the sizes that do a relevant division of powders according to particle sizes. In view of very small mesh size and mass of particles of nanopowders the known methods of division either are applicable restrictedly or are not applicable absolutely.

In 1787 Chladni showed that at the fluctuation of a plate a bow of a particle of sand self-organizing symmetric structures form. It was also noticed that more shallow particles of sand accumulate on a crest of fluctuations of a plate. The physical reasons for particle size distributions are bound to streams of air or other environments which is over a plate. Recently there was interest in these streams within the creation of minifans in the form of the vibrating plates of the infinite and finite lengths for microelectronic devices, so, models of gas streams are received in work [1]. In work [2] the idea of Chladni is applied in nanomechanics to a division of nanoparticles according to the sizes by acoustic waves of different frequencies, however for very high frequencies and small sizes of a plate that is almost inapplicable.

In this work the possibility of division of nanoparticles on a plate of a rather big size and with low frequencies is shown, and on the basis of the method of successive approximations [3] of the Navier-Stokes equations for a vorticity the equations and boundary conditions for fluctuating plates taking into account amplitude to the second-order are received. Through the Mathematica system analytical decisions are received and the streams of vectors fields of speeds for a plate of the infinite length proving particle sorting by an acoustic method are constructed.

2. The experimental part

As in our case practical division of nanopowders according to particle sizes matters, as the fluctuating plane the beam (cantilever) length of which much more than its width is used. Since one end the cantilever is excited by transverse vibrations with his resonant frequency 1,035 kHz. On a surface of a cantilever nanopowder of dioxide of silicon, which was redistributed to nodal lines and in the form of clouds in crests (Figure 1), is sprayed.

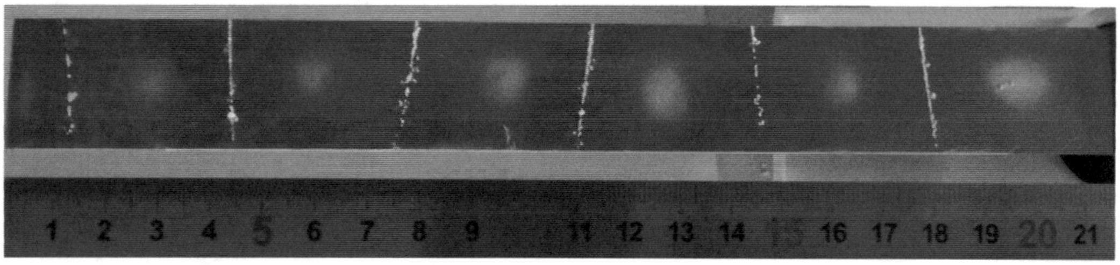

Figure 1. The fluctuating cantilever with silicon dioxide nanopowder on the surface

In figure 2 distributions by the sizes of the powder which is selected from nodal lines and from crests, the light scattering received by the method is represented. The experiment series of distributions by the sizes was carried out. The average size of powder was of nodal lines 65μm (so big size is caused by the agglomeration of particles), in crests – 3.5 μm. The division of powder according to the sizes is obvious.

It is expedient to make the calculation of vector fields of gas streams over a surface of this cantilever in view of the fact that the division of powder according to particle size on it is confirmed experimentally.

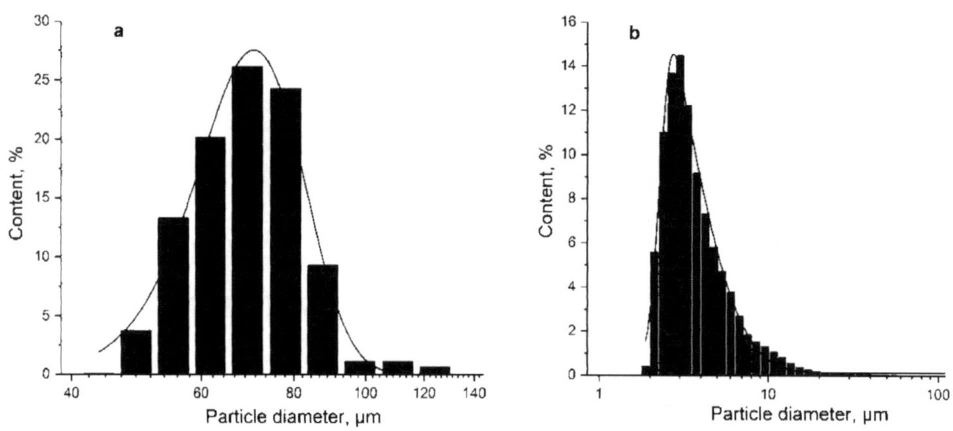

Figure 2. Distribution by the nanopowder sizes from the nodal line (a) and from a crest (b)

3. Numerical modelling of gas streams over a plate

We will consider a plate vibrating in the longitudinal direction under the law $y = w(x,t)$ with amplitude A<<1 and harmonic frequency σ.

In a two-dimensional flow of viscous incompressible fluid, the vector of vorticity has only one nonzero component:

$$\omega_z = \omega = \frac{\partial U}{\partial x} - \frac{\partial V}{\partial y} = -\frac{\partial^2 \psi}{\partial x^2} - \frac{\partial^2 \psi}{\partial y^2} \equiv -\nabla^2 \psi \ ,$$

where flow function $\psi(x, y)$ is defined as follows:

$$U = \frac{\partial \psi}{\partial y}, \quad V = -\frac{\partial \psi}{\partial x}.$$

Further from Navier-Stokes's equations it is received the equation for a vorticity:

$$\frac{\partial \omega}{\partial t} + U\frac{\partial \omega}{\partial x} + V\frac{\partial \omega}{\partial y} = \nu\nabla^2 \omega, \tag{1}$$

where ν – kinematic viscosity.

Using the method of successive approximations [3] we decompose flow function:

$$\psi(x, y, t) = \psi_1(x, y, t) + \psi_2(x, y, t) + \psi_3(x, y, t) + \ldots, \tag{2}$$

where $\psi_i = O(A^i), \quad i = 1, 2, \ldots$

Substituting (2) in (1) and equating members of an identical order on amplitude to the second order we receive:

$$\begin{cases} \dfrac{\partial \nabla^2 \psi_1}{\partial t} = \nu\nabla^4 \psi_1, \\ F(x, y, t) + \dfrac{\partial \nabla^2 \psi_2}{\partial t} = \nu\nabla^4 \psi_2, \end{cases} \tag{3}$$

where $F(x, y, t) \equiv \dfrac{\partial \psi_1}{\partial y}\dfrac{\partial \nabla^2 \psi_1}{\partial x} - \dfrac{\partial \psi_1}{\partial x}\dfrac{\partial \nabla^2 \psi_1}{\partial y}$

From sticking conditions on a plate we receive nonlinear boundary conditions at vibrating plate $y = w(x, t)$:

$$U = \frac{\partial \psi}{\partial y} = 0, \quad V = -\frac{\partial \psi}{\partial x} = \frac{\partial w}{\partial t} \tag{4}$$

Considering smallness of amplitude we decompose (4) in Taylor row in case of fixed plate $y = 0$:

$$\frac{\partial \psi}{\partial y}(x, w, t) = \frac{\partial \psi}{\partial y}(x, 0, t) + w\frac{\partial^2 \psi}{\partial y^2}(x, 0, t) + \frac{w^2}{2!}\frac{\partial^3 \psi}{\partial y^3}(x, 0, t) + O(A^4)$$

$$\frac{\partial \psi}{\partial x}(x, w, t) = \frac{\partial \psi}{\partial x}(x, 0, t) + w\frac{\partial^2 \psi}{\partial x \partial y}(x, 0, t) + \frac{w^2}{2!}\frac{\partial^3 \psi}{\partial x \partial y^2}(x, 0, t) + O(A^4)$$

Substituting further in (4) and equating identical members on amplitude we receive:

$$\begin{aligned} &\frac{\partial \psi_1}{\partial y}(x, 0, t) = 0, \quad \frac{\partial \psi_1}{\partial x}(x, 0, t) = -\frac{\partial w}{\partial t}, \\ &\frac{\partial \psi_2}{\partial y}(x, 0, t) = -w(x, t)\frac{\partial^2 \psi_1}{\partial y^2}(x, 0, t), \quad \frac{\partial \psi_2}{\partial x}(x, 0, t) = -w(x, t)\frac{\partial^2 \psi_1}{\partial x \partial y}(x, 0, t). \end{aligned} \tag{5}$$

From conditions of attenuation of fluctuations of fluid at infinity it is in addition received, as their gradients tend to zero at $y \to \infty, \quad x \to \pm\infty$:

$$\begin{aligned} &\psi_1(x, \infty, t) = \frac{\partial \psi_1}{\partial x}(x, \infty, t) = \psi_2(x, \infty, t) = \frac{\partial \psi_2}{\partial x}(x, \infty, t) = 0, \\ &\psi_1(\pm\infty, y, t) = \frac{\partial \psi_1}{\partial x}(\pm\infty, y, t) = \psi_2(\pm\infty, y, t) = \frac{\partial \psi_2}{\partial x}(\pm\infty, y, t) = 0. \end{aligned} \tag{6}$$

From the recurrent system of boundary value problems (3), (5), (6) received above were calculated analytically in the Mathematica system by method of variables separation of flow function (2) stationary currents to the second order of expansion on amplitude.

Vector fields of flow functions of the first and second order of expansion on amplitude respectively are given in Figures 3-4.

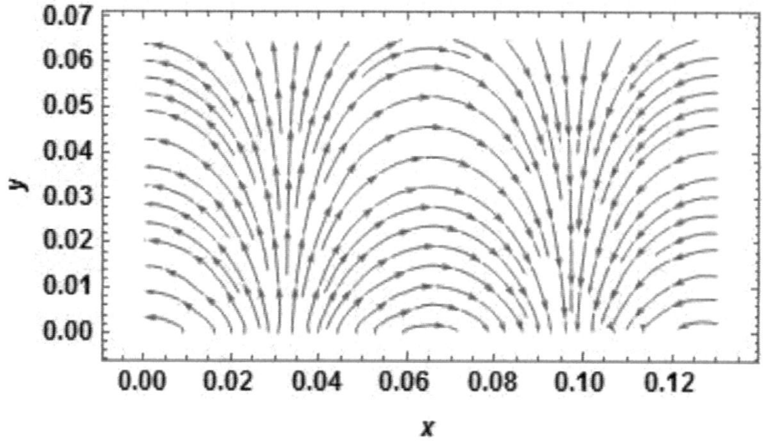

Figure 3. Vector field of flow function $\psi_1(x, y)$

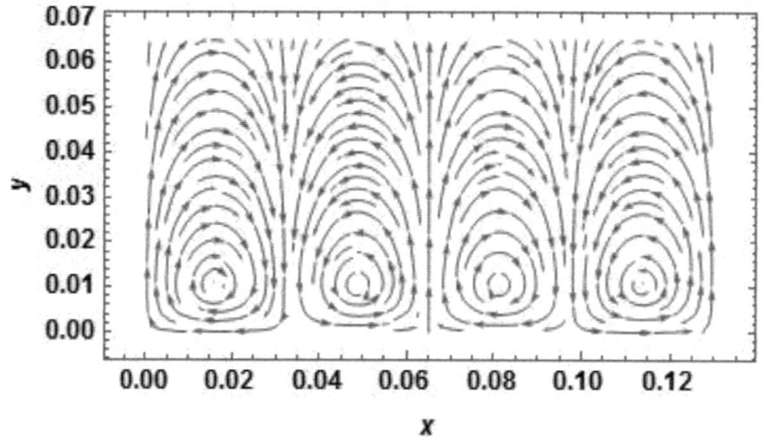

Figure 4. Vector field of flow function $\psi_2(x, y)$

In calculations the following data were used:

$$w(x,t) = A\cos\left(\frac{\pi}{L}x\right)e^{j\sigma t},\ j=\sqrt{-1}, \nu = 1.45x10^{-5}m^2/s,\ L=6.5cm,\ A=0.08cm,\ \sigma=120\pi\ rad/s.$$

4. Conclusion

It is experimentally shown that the division of nanoparticles according to the sizes into surfaces of the fluctuating plate is possible and with a frequency of about 1 kHz.

Numerical modelling taking into account decomposition on amplitude to the second-order shows (figure 4) that streams over the fluctuating plate have vortex structure with the period equal L/2. The next whirlwinds rotate in mutually opposite directions with the centers near a plate. Outflow comes from a plate in the neighborhood of nodal points, and dribble to a plate in the neighborhood of crests of a standing wave where particles of shallow fraction in the form of round clouds (figure 1, figure 2) accumulate.

References

[1] Acikalin T, Ramah A and Garimella S V 2003, Two-dimensional streaming flows induced by resonating thin beams, *J.Acoust.Soc. Am.* **114** (4.1) 1785-95.

[2] Dorrestijn M 2007 Chladni Figures Revisited based on nanomechanics, *Physical Review Letters* **98** 026102

[3] Sclichting H 1955, *Boundary layer theory*, (New-York: McGraw-Hill)

IOP Conf. Series: Materials Science and Engineering **704** (2019) 012015 doi:10.1088/1757-899X/704/1/012015

Dispersed-colloidal fuel systems

M Baranova[1], I Grishina[2], B Damdinov[2], R Gomboev[2]
[1] Krasnoyarsk SAU, Krasnoyarsk, Russia,
[2] Siberian Federal University, Krasnoyarsk, Russia

E-mail: marina60@mail.ru

Abstract. A substantiation of the action mechanism of alkaline additives on the rheological characteristics of binary fuel systems is proposed, which will allow us to proceed to numerical methods for studying the technological processes of production. The possibility of increasing the stability of the suspension by using a nanomodifier obtained by ultradispersing to a submicron size of a portion of the same suspension is shown.
Key words: rheology, fuel system, numerical methods, nanodispersions

Getting fuel water-coal suspensions from coals of different degrees of metamorphism and from waste of coal-enrichment and coal processing is still relevant, especially for coal regions, because solid-fuel energy generation leads to global pollution of the environment [1]. The transfer of coal and coal sludges to water-coal suspension fuel makes it possible to improve the situation when burning coal in this form [2]. Essentially fuel water-coal suspensions are binary systems in which the solid phase should have the smallest possible size and maximum fluidity with a minimum content of the carrier medium. For this, plasticizer reagents are added to the system, also in a minimal amount, up to 1% for a dry weight of coal [3].

In the course of active mechanical and chemical destruction of coal particles in the aqueous phase, a binary system is formed. This system is called a suspension, but the carrier medium is colloidal solutions of nanoparticles in a liquid solvent. Due to the small size of the inclusions, such systems have special physical and chemical properties. Systems have an increased surface energy due to a large number of atoms in an excited state and having at least one free electron at an external energy level. Such systems may well be attributed to nanodispersions. Because of its structure and instability in the size of aggregates of nanoparticles, nanodispersions, as a rule, are rather unstable. Their properties are easily changed and depend on external influences. The main task that must be solved on the way to their industrial use is to obtain stable nanodispersions with reproducible properties. Thus, the technological and thermal characteristics of fuel water-coal suspensions intended for direct combustion are largely determined by the degree of grinding and the physical and chemical composition of the used coals [4].

The purpose of this work was to develop theoretical foundations for technological solutions for the production, transport and combustion of water-coal suspensions from brown coal.

To ensure that the fuel can be stored, transported through pipes, fed for combustion and burned in furnaces, it is necessary to comply with the requirements for rheological characteristics [5] and stability. It is known that mineral impurities contained in the solid phase contribute to the formation of coagulation structures in the WUS, thereby affecting on rheological properties and sedimentation resistance. During the work a number of experiments were carried out to study the effect of the amount of mineral impurities on the viscosity and stability of suspensions. For this purpose, brown coals of grade B2 were used with an ash content of 10.6 to 34.7%. From each coal sample with a certain ash content, a series of suspensions with different mass fraction of the solid phase were obtained. The content of the solid phase of the obtained suspensions without the use of plasticizing additives was, on the average, 43-46%, depending on the ash content. The lowest heat of WCS combustion in this case is 5.61-7.66 MJ / kg.

To increase the solid content in the WCS, a number of plasticizing additives having an alkaline medium, such as technical lignosulfonates of KBP, Lifrin-P, Ligrin, Lipor, were studied. It was found that as the ash content of the initial coal increased, the diluting effect of the additive decreased, which led to the need to increase the amount of plasticizer or to use the complex composition of the additive.

But the positive point, in this case, was that, unlike sodium hydroxide, when using lignosulfonate there is no reversion of the change in structural viscosity. This circumstance is extremely important, since the introduction of KBP in industrial conditions exceeding the optimal dosages (1%) will not cause a deterioration in the fluidity of the suspensions. In the case of using *NaOH*, strict regulation of its concentration (0.2-0.5%) is required. This concentration for alkali is critical, since exceeding the indicated concentration, the structural viscosity of the WCS increases, also there is the formation and precipitation of a dense precipitate. The noted properties indicate the possibility of optimizing the parameters of suspensions under industrial conditions by a joint change in the amount of the additive and by regulating the ash content of the coal [6-8]. It was found that the effective dilution of the water-coal system depends on the content of native humic acids in brown or oxidized coal.

The nature of the alkalis effect on the stability of coal-water suspensions based on brown coal is practically not investigated and is not discussed in the literature. Understanding the mechanism of action of alkaline additives on the rheological characteristics of binary fuel systems will ultimately allow us to avoid the costly experiments and switch to numerical methods for studying the technological processes of production. The dependence of the structural viscosity of the WCS on the concentration of the introduced alkali has an extreme character with a minimum in the range of alkali concentrations from 0.2 to 0.5%. With the increase in coal ash content from 5 to 8%, the amount of sodium hydroxide needed to be increased at least twice. But an increase in the content of alkali to 1% led to the stratification of the suspension to form a solid precipitate. The following mechanism is suggested for the impact of alkali and its concentration on the viscosity of the WCS.

Water alkali, reacting with brown coal, in the first stage, transfers to the aqueous phase the humic acids in the form of sodium salts:

$$\mathrm{NaOH} + (\mathrm{HO\text{-}Humin})_{\mathrm{coal}} \rightarrow (\mathrm{NaO\text{-}Humin})_{\mathrm{aque.}} + \mathrm{H_2O} \quad (1)$$

where $(\mathrm{HO\text{-}Humin})_{\mathrm{coal}}$ – humic acids in coal, $(\mathrm{NaO\text{-}Humin})_{\mathrm{aque.}}$ – dissolved sodium salts of humic acids. The sodium salts of humic acids in the aqueous solution are in the form of a true solution, and this composition has an aqueous phase of the WCS at high concentrations of alkali (more than 0.5%), providing a complete transition of humic acids to the corresponding salts.

At low concentrations of alkali, the first stage of its interaction with coal is described by equation (1), followed by the stage of hydrolysis of salts of humic acids and migration of sodium ions from solution to the solid phase of coal:

$$(\mathrm{NaO\text{-}Humin})_{\mathrm{aque.}} + (\mathrm{HO\text{-}Humin})_{\mathrm{coal}} \rightarrow (\mathrm{HO\text{-}Humin})_{\mathrm{gel}} + (\mathrm{NaO\text{-}Humin})_{\mathrm{coal}} \quad (2)$$

Process (2) can proceed precisely at low concentrations of alkali, and this will lead to coagulation of humic acids, the transition of a true solution of sodium salts of humic acids to a colloidal solution, a gel of humic acids. The thus formed gel stabilizes the water-coal suspension and reduces its viscosity. At high concentrations of alkali, process (2) is impossible, because all acid groups of humins are deprotonated, and this leads to precipitation of hard sediments with an alkali concentration of about 1% [9].

The observed regularities demonstrate the possibility of optimizing the properties of suspensions under industrial conditions by a joint change in the amount of the additive and by regulating the physical and chemical composition of the coal.

Water-coal fuel is classified as ballasted and its burning requires a special approach. Nanoactivators and nanomodifiers are used to improve physical-chemical and physico-mechanical characteristics of dispersed-colloidal systems to control their structural-rheological and thermophysical characteristics, as well as the process of ignition and stability of the combustion process. Nanoactivators are organic compounds of cluster type - nanoparticles (20 nm), consisting of several identical molecules. They have a secondary structure due to internal hydrogen bonds, which provide the activation effect at relatively low temperatures (40°C), since hydrogen bond protons have very low rotational-vibrational excitation levels. The nanomodifier was obtained from a sample of the basic composition of the suspension, ground to submicron sizes. The addition of 1% modifier increases the stability of the suspension by approximately 100% (with the same amount of stabilizing chemical additives).

The developed nanoactivators have a very strong overall structure, due to which they are able to activate flue gases at high temperatures. It should be noted that there are problems with determining the percentage of injected nanoactivators. If the concentration of the nanoactivator in the activated liquid medium is higher than a certain limiting level, then the activation process does not arise, because the internal energy of the medium is not sufficient to transfer more particles of the nanoactivator into a pseudostable excited state. With insufficient concentration of the nanoactivator, the intensity of the coherent radiation is not sufficient to activate the medium at multiple frequencies. Therefore, for different media and conditions, there are two concentration levels - the lower and upper, within which the process is stable. The activation process takes place in the combustion chamber and spreads to the combustion gases. For coal-fired boilers, an aqueous solution of the nanoactivator is used, which is injected into the combustion chamber. Specific fuel consumption is reduced by 5-7% with a nanoactivator consumption of about 0.5 g / t coal.

Conclusion

The influence of the amount of ash component of brown coal and native humic acids on the rheological characteristics and static stability of water-coal fuel suspensions is established.

A substantiation of the action mechanism of alkaline additives on the rheological characteristics of binary fuel systems is proposed, which will allow us to proceed to numerical methods for studying the technological processes of production.

The possibility of increasing the stability of the suspension by using a nanomodifier obtained by ultradispersing to a submicron size of a portion of the same suspension is shown. The addition of up to 1% of the modifier increases the suspension stability on 50-100%. Specific fuel consumption is reduced by 5-7% with a nanoactivator consumption of about 0.5 g / t coal.

Supported by RFBR 18-02-00523a.

References

[1] Murko V, Baranova M and Grishina I 2019 *J. Phys.: Conf. Series.* **1261**(1) 012024,

[2] Baranova M, Li Q, Zheng Zh Y, Li F Ch, Kulagin V A and Likhachev D S 2014 *J. of Sib. Fed. Univ.: Technics and technol-s.* **7** (4) 474-9,

[3] Leong Y K, Creasy D E, Boger D V and Nguyen Q D 1987 *Rheol. Acta.* **26** (3) 291,

[4] Baranova M and Grishina I 2011 *Polzunovsky Vestnik.* **2/1** 235,

[5] Damdinov B, Dembelova T, Badmaev B and Barnakov Y 2019 *Solid State Phen.* **288** 130,

[6] Murko V, Kulagin V and Baranova M 2017 *J. of Sib. Fed.Univ.: Engin. and technol-s* **10 (8)** 985-92,

[7] Musalam A 2016 *Intern. J. of Energy and Envir. Res.* **4**(3) 27,

[8] Murko V I, Fedyaev V I, Karpenok V I, Zasypkin I M, Senchurova Y A and Riesterer A 2015 *Thermal Science.* **19**(1) 243,

[9] Baranova M P , Kulagin V A and Taraban'ko V E 2011 *Rus. J. of Appl. Chem.* **84**(6) 939-44.

A method to determine uranium compounds in drinking water

N Tegshbayar[1*], Kh Tsookhuu[1], O Bolormaa[2]
[1]Department of Physics, National University of Mongolia, Ulaanbaatar-14201, Mongolia
[2]Departmnet of Chemistry, National University of Mongolia, Ulaanbaatar-142018 Mongolia

e-mail: infomax999@gmail.com

Abstract.
In this paper we have studied the possibility of determining the forms of uranium ions and minerals dissolved in water by using the Pourbaix diagram. Uranium in water with low uranium content exists in the form of uraninite or uranium dioxide UO2+ and with higher uranium content in the form of ions $UO_2(CO_3)$. By selecting carbonate minerals and calculating the saturation index, it is shown that uranium in the water of Eastern region of Mongolia dissolved in the forms of minerals such as quartzite $CaMgUO_2(CO_3)_3(H_2O)_{12}$, andersonite ($Na_2CaUO_2(CO_3)_3(H_2O)_6$), bayleyite ($Mg_2UO_2(CO_3)_3(H_2O)_{18}$), liebigite ($Ca_2UO_2(CO_3)_3(H_2O)_{10}$).
Keywords: water, uranium, correlation, Piper, Pourbaix diagram, uranium minerals

Typical spectroscopic and chemical analysis techniques determine only the elemental compositions of the substances. However, the structural information of the chemical compound is determined by the molecular spectroscopic methods such as Raman spectroscopy and slow neutron scattering. In this paper, we discuss about a non-instrumental method to determine uranium compounds in water. The method is known as a Pourbaix diagram that maps out possible equilibrium states of a thermodynamic system consisting of ions, molecules, crystals and, metals in the solvent. Pourbaix diagram shows a dependence between oxidation and reduction potential (ORP or E-h) and the water pH [1]. In other words, it shows the possible equilibrium phases of a system in the given pH and oxidation and reduction values. The diagram can be drawn for each indiviual elements. However, it depends on temperature, pressure metal binding agent (ligands) concentration in the solution that correlated by Nernst equation. Ligands are determined by general physico-chemical characteristics and major-ion analysis of the water. General physico-chemical characteristics of water includes pH, temperature, electrical conductivity (EC), oxidation and reduction potential (ORP), total dissolved solids (TDS), dissolved oxygen (DO). The main ion analysis can effectively be shown with Durov and Pipers diagram [2].

The Pourbaix diagram shows only the thermodynamical equilibrium states of the solutions, yet has no information of the metastable states. To estimate the metastable forms of uranium in water, we took into account the hydrolysis of 4 valence uranium ion and metastale bonds between uranyl ion, UO_2^{2+}, and carbonate, chloride and sulfates. Moreover, we calculated uranium-ion correlation [3,4].

We studied groundwater samples of Dornogovi (DG) and Sukhbaatar (SB) aimag where uranium content is relatively high. The results of the measurements are shown in the figures 1-4. The diagram indicates the water samples of Dornogovi aimag is more alkaline and for Sukhbaatar aimag it is more acidic (Fig.1).

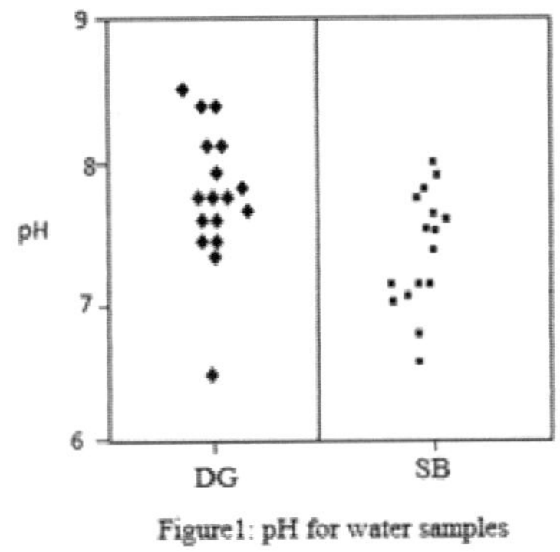

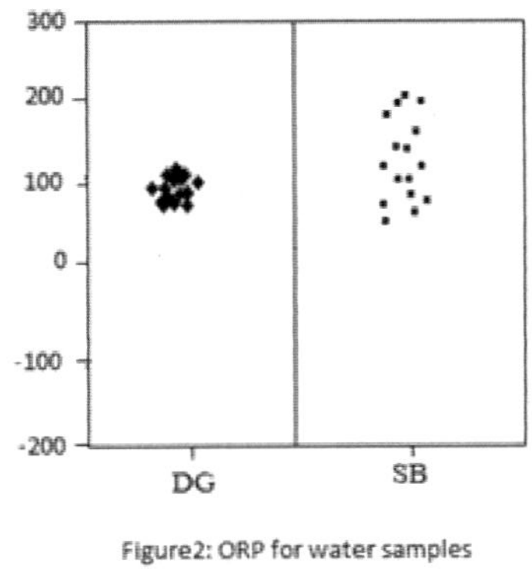

Figure 1: Water pH for samples of Dornogovi (DG) and Sukhbaatar (SB) aimags.
Figure 2: Oxidation and reduction potential (E-h) for samples of Dornogovi and Sukhbaatar aimags

From the diagram we see that E-h scattering range for the samples of Dornogovi aimag is narrow and while it is wider for samples of Sukhbaatar aimag [4].
Figures 3 and 4 show the results of major ions in water using by Durov diagram.

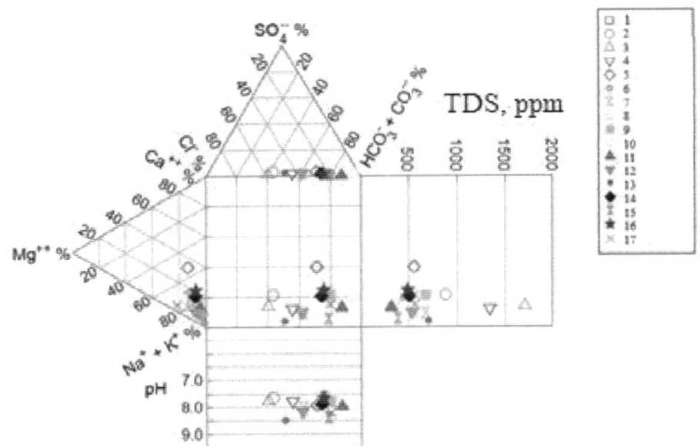

Figure 3: Durov diagram for water samples of Dornogovi aimag

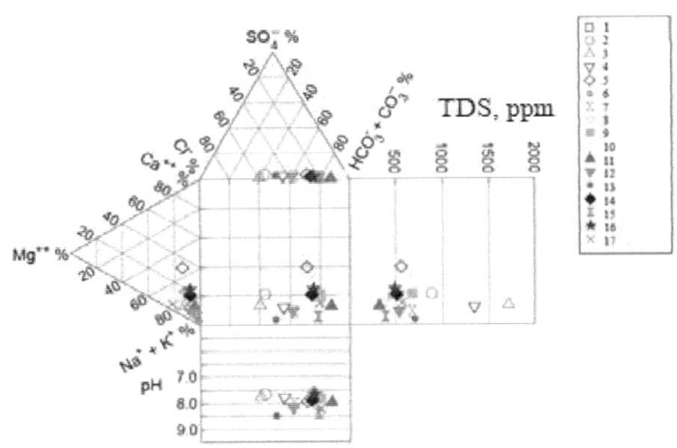

Figure 4: Durov diagram for water samples of Sukhbaatar aimag

From the diagrams it illustrates that ion content of the water samples of two aimags are almost same and Na and K ions among the cations and hydrocarbobate ions among the anions prevail. In the Tables 1 to 4 the correlation coefficient for uranium- physicochemical characteristics and uranium – major ions are showed for two selected images.

IOP Conf. Series: Materials Science and Engineering **704** (2019) 012016 doi:10.1088/1757-899X/704/1/012016

Table 1: Correlation for uranium and physico-chemical characteristics of water samples of Dornogovi aimag

	U	pH	EC	TDS	ORP	DO
U	1					
pH	0.02	1				
EC	0.136	0.14	1			
TDS	0.111	0.118	0.968	1		
ORP	-0.168	-0.134	0.205	0.408	1	
DO	0.056	0.078	-0.156	-0.313	-0.654	1

$r_{xy}(0.05:24)=0.381$

Table 2: Correlation for uranium and major ions of water samples of Dornogovi aimag.

	U	pH	EC	TDS	ORP	DO
U	1					
pH	0.02	1				
EC	0.136	0.14	1			
TDS	0.111	0.118	0.968	1		
ORP	-0.168	-0.134	0.205	0.408	1	
DO	0.056	0.078	-0.156	-0.313	-0.654	1

$r_{xy}(0.05:24)=0.381$

Table 3: Correlation for uranium and physico-chemical characteristics of water samples of Sukhbaatar aimag

	pH	EC	TDS	ORP	Do	Uranium
pH	1					
EC	0.096	1				
TDS	0.092	0.999	1			
ORP	0.033	-0.195	-0.199	1		
DO	0.021	-0.050	-0.034	-0.021	1	
Uranium	-0.209	0.483	0.490	-0.458	-0.137	1

$r_{xy} = (0.05, 14) = 0.480$

The tables show that uranium in water of Dornogovi aimag has not any correlation with physico-chemical characteristics and major ions. However, uranium in water of Sukhbaatar aimag is positively correlated with EC, TDS and weakly negative correlated with E-h. Also, uraniumin in water positively correlated with hydrocarbonate and chloronium ions in case of Sukhbaatar. Based on the above mentioned remarks we have built a Pourbaix diagram for 2 samples, one of these with low uranium content (2ppb) and second is higher uranium content (212ppb). Pourbaix diagrams are shown in Figure 5 and 6. Crossed perpendicular line in the diagrams show corresponding value of pH and E-h for selected samples. The Pourbaix diagram show that the uranium in water, with low uranium content can be exist in the form of uraninite (UO2) and the uranium in water, with higher uranium content exist in the form of $UO_2(CO_3)_2^{2}$ ions.

Table 4: Correlation for uranium and major ions of water samples of Sukhbaatar aimag.

	HCO_3^-	Hardness	Cl^-	SO_4^{2-}	NH_4^+	NO_2^-	NO_3^-	Uranium
HCO_3^-	1							
Hardness	0.481	1						
Cl^-	0.626	0.658	1					
SO_4^{2-}	0.558	0.599	0.922	1				
NH_4^+	0.398	0.713	0.576	0.403	1			
NO_2^-	-0.549	-0.096	-0.186	-0.138	-0.397	1		
NO_3^-	0.147	0.385	0.268	0.188	0.347	-0.031	1	
Uranium	0.634	0.562	0.480	0.452	0.232	-0.329	-0.074	1

r_{xy} = (0.05, 14) = 0.480

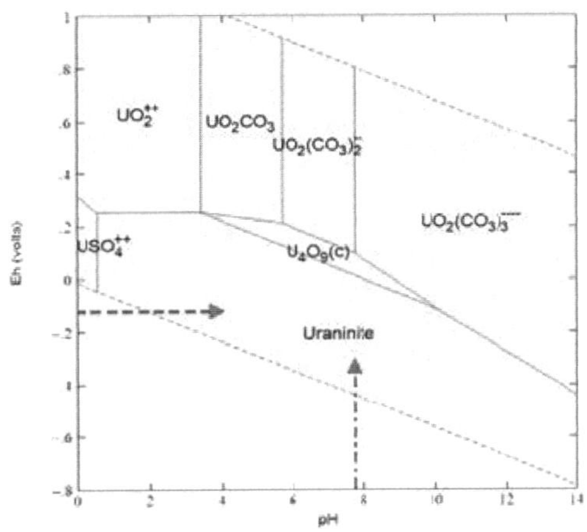

Figure 5: Pourbaix diagram of the sample with low uranium content

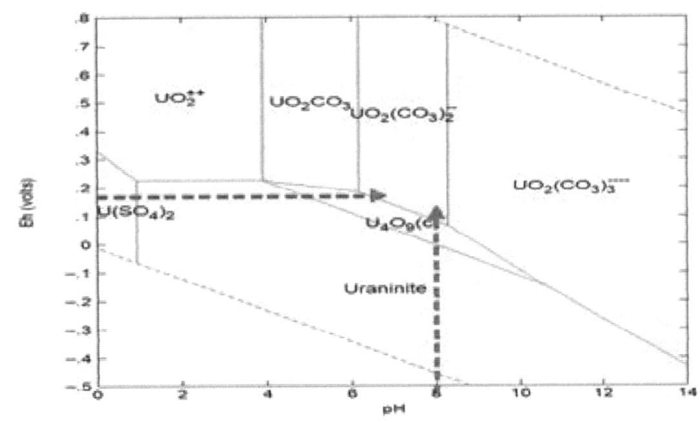

Figure 6: Pourbaix diagram of the sample with higher uranium content

Further we studied which cations can be reacted with $UO_2(CO_3)_2^{2}$ uranium ions dissolved in the water. Saturation index was calculated using Geochemist Workbench program[6]. By selecting carbonate minerals and calculating the saturation index we have predicted that the uranium in water can be dissolved in the forms of quartzite $CaMgUO_2(CO_3)_3(H_2O)_{12}$, andersonite ($Na_2CaUO_2(CO_3)_3(H_2O)_6$), bayleyite ($Mg_2UO_2(CO_3)_3(H_2O)_{18}$) and liebigite ($Ca_2UO_2(CO_3)_3(H_2O)_{10}$).

References

[1] Pourbaix M 1974 Atlas of electrochemical equilibria in aqueous solutions. 2d English ed., Houston, Tex.: National Association of Corrosion Engineers.

[2] Piper A M 1953 A Graphic Procedure in the Geochemical Interpretation of Water Analysis. Washington D.C, United States Geological Survey, OCLC 37707555, ASIN B 0007HRZ36

[3] Ravikumar P, Somashekar R K and Prakash K L 2015 A comparative study on usage of Durov and Piper diagrams to interpret hydrochemical processes in groundwater from SRLIS river basin, Karnataka, India, *Elixir Earth Sci.* **80** 31073-7

[4] Tsookhuu Kh, Bolormaa O, Tegshbayar N 2017 Uranium concentration in drinking waters of Mongolia *NUM J. Science* **25**(478) 116-32.

[5] Dougles G, and Brookins 1987 Eh-pH diagram for Geochemistry. USA 145-50.

[6] Iwatsuki T, and Arthur R 2004 Solubility constraints on uranium concentrtion in groundwaters of Tono uranium deposit, *Japan. Radiochimica Acta* **92**, 792-793.

Numeric Modeling of Phononic Crystal with Time-Dependent Properties

A Demin[1,a], B Damdinov[2,3,b], Y Baloshin[4,c]

[1]Buryat State University, Smolina 24a, Ulan-Ude, Russia, 670000
[2]Siberian Federal University, Svobodny 79, Krasnoyarsk, Russia, 660041
[3]Institute of Physical Materials Science of SB of RAS, Sakhyanovoy 6, Ulan-Ude, Russia, 670047
[3]ITMO University, Kadetskaya line 3b, Saint-Petersburg, Russia, 199034

[a]tonydomi@mail.ru, [b]dababa@mail.ru, [c]baloshin@mail.ru

Abstract. The research is devoted to numeric modeling of phononic crystals with time-depended properties of periodic structure. There are many works describing phononic crystal with a tunable bandgap. The mechanism of restructuring can have a different physical nature. In this turn, we developed model of phonon crystal with fast-changing properties, where frequency of changing and propagating wave have the same order. The influence of time-dependent characteristics on propagation of wave packets, waves with different initial phase is shown in this work. Parameters of the crystal cell are given by time-dependent; signal propagation has significant differences from the static case. It was shown that under certain conditions a phononic crystal acquires the properties of a phase filter.
Keywords: modeling, wave propagation, material properties, phononic crystal, structure

Introduction

A phononic crystal is an artificial crystal of a finite-size periodic array composed of elements with different acoustic impedance. It should have full band-gaps, or complete band-gaps, where any sound wave is not allowed to propagate into the crystal but is reflected completely by the crystal [1, 2]. Several classes of phononic crystals materials differing by the physical nature of the inclusions and the matrix have been studied. By the nature of the materials they are solid/solid, fluid/fluid, and mixed solid/fluid composite systems. In the considered case, solid/solid material only longitudinal modes are allowed. There are opportunities of shear waves in suspension and viscoelastic materials.

Materials with tunable components are interesting for us. Rebuilding is possible by controlling the electronics, which can change material properties both independently of the transmitted signal and with feedback dependent on the signal [3, 4]. Properties changing of the material by influence to the medium by an external field were also investigated, for example, in work [5]. A number of studies have investigated the change in the speed of sound in materials located in magnetic and electric fields. For magnetic liquids in a magnetic field, the change can reach 50%. In our model, the change in sound velocity occurs due to a 10% change in the bulk modulus, which corresponds to a change in the sound velocity by approximately 3% [6].

Method

For simulation, we used finite-difference time-domain (FDTD) method. The FDTD method has recently become popular for acoustics simulation. FDTD solves acoustical wave field in medium, by employing finite differential equations to approximate the derivatives of pressure and the particle

IOP Conf. Series: Materials Science and Engineering **704** (2019) 012017 doi:10.1088/1757-899X/704/1/012017

velocity, both temporally and spatially. The use of the FDTD method allows solving problems with time-dependent parameters of the media [7].

Sound propagation in the fluid is described by two kinds of differential equations, Euler's equation and the equation of continuity. For the loss media case [8]:

$$\rho\frac{\partial u}{\partial t}+\frac{\partial P}{\partial x}+\alpha_x u=0\,,$$
$$\rho\frac{\partial v}{\partial t}+\frac{\partial P}{\partial y}+\alpha_y v=0\,,$$
$$\rho\frac{\partial \omega}{\partial t}+\frac{\partial P}{\partial z}+\alpha_z \omega=0\,,$$
$$\frac{\partial P_x}{\partial t}+K\frac{\partial u}{\partial x}+\rho\beta_x\rho P_x=0\,,$$
$$\frac{\partial P_y}{\partial t}+K\frac{\partial v}{\partial y}+\rho\beta_y\rho P_y=0\,,$$
$$\frac{\partial P_z}{\partial t}+K\frac{\partial \omega}{\partial z}+\rho\beta_z\rho P_z=0$$
$$P=P_x+P_y+P_z\,,$$

where P_x, P_y, P_z - are components of pressure, *u, v, w* - are particle velocity in x,y,z direction, respectively, *K*, ρ - volume elastic ratio and density and α, β - the factors responsible for sound absorption. The discretization of similar system is described in detail in [9].

Result

The model shown in the figure 1 is an acoustical waveguide of variable cross section. As can be seen in the figure, a port on the left is a source of acoustic waves, and on the right, respectively, is a receiver. The ends are absorbent and implemented as perfect matched layers. Other part of waveguide wall is rigid and velocity is equal zero in this wall.

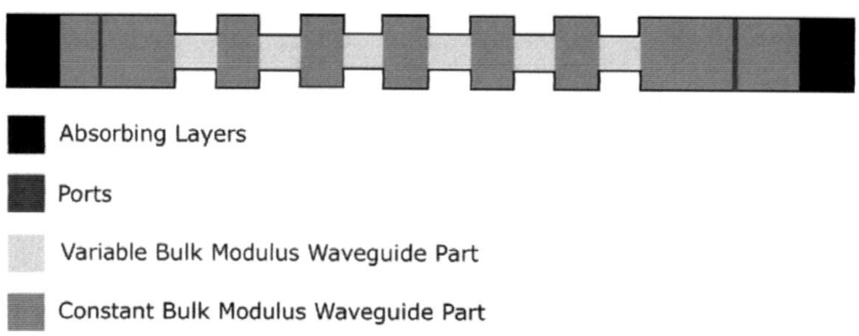

Fig. 1. General view of a numerical experiment

At the first stage, we calculated the dependence of pressure amplitude on frequency for the cases with a bulk modulus of K = 2 GPa and K = 2,2 GPa, with the same density 1000 kg/m^3 The result of the calculation is shown in the figure. Different wave propagation is observed in the 22 kHz region.

We choose the frequency of 21750 Hz as the frequency at which the tuning is possible. Further the calculation was made for a crystal with time-dependent properties. Bulk modulus changes with a frequency four times the frequency of the signal from 2 GPa to 2,2 GPa for indicated on figure 2 region.

IOP Conf. Series: Materials Science and Engineering **704** (2019) 012017 doi:10.1088/1757-899X/704/1/012017

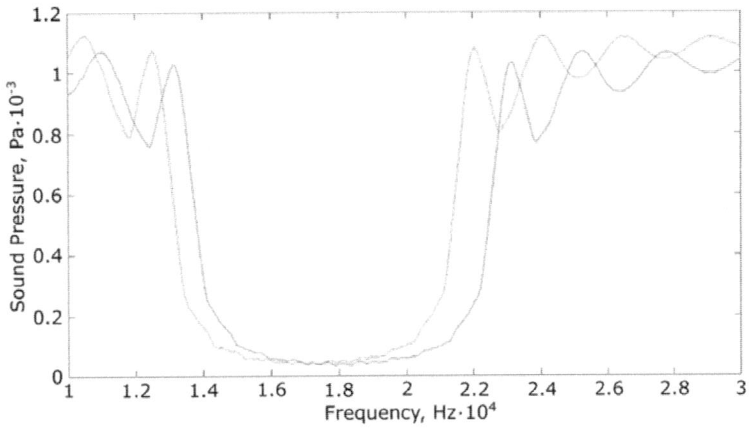

Fig. 2. Bandwidths for $K = 2{,}2$ (red) and $K = 2$ GPa (blue)

This variation leads to new properties of phononic crystal. Initial phase of signal have influence on signal transition. This dependence is shown in figure 3.

As you can see in figure 4 signal transition depends on phase. In this way the attenuation of acoustic wave with equal frequency but different phase can appear in time-dependent phonon crystal.
This effect is provided by tuning of the properties at various period of signal transition causes changes in reflection coefficients on the elements of the crystal

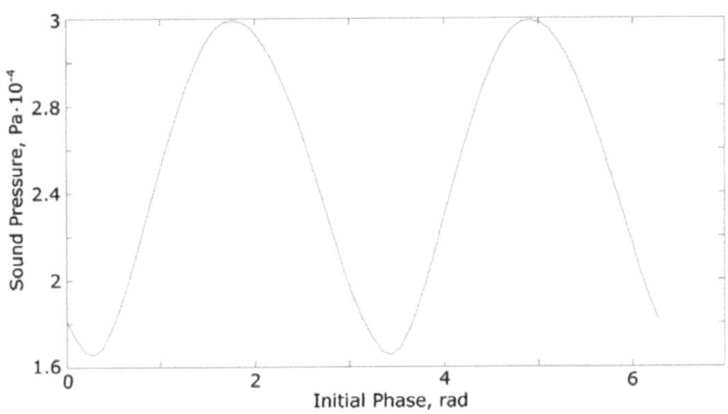

Fig. 3. Dependence of the pressure amplitude on the initial phase.
Frequency of signal is 21750 Hz, frequency of modulation is 87 kHz

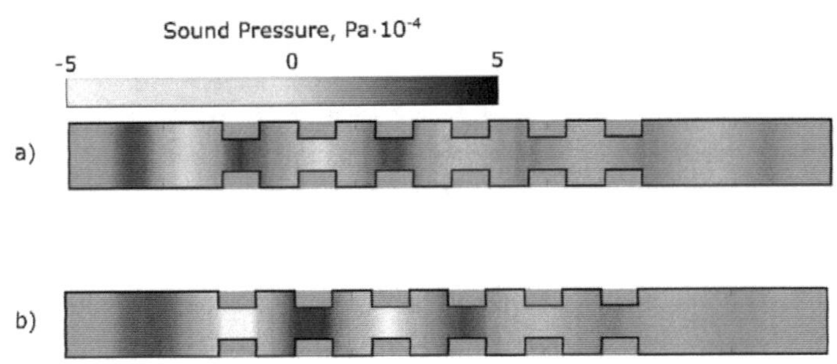

Fig. 4. Comparison of pressure distribution in different phase ϕ.
a) $\phi = 1{:}25$ radians, b) $\phi = 0{:}25$ radians

Summary
In summary, we investigated the models of phononic crystal with time-dependent properties. In our numeric approach the changes of crystal properties occurs due to media properties variation. Supposed the turning mechanism may be changes of ferromagnetic liquid in magnetic field. It was shown that the modulation of media of a phononic crystal lead to dependence of signal amplitude on initial phase. It is possible to expand this model to 2-D and 3-D case.

Supported by RFBR grant №18-02-00523.

References

[1] Sigalas M and Economou E N 1993 Band structure of elastic waves in two dimensional systems. *Solid State Commun.* **86**(3) 141–3.
[2] Kushwaha M S, Halevi P and Djafari-Rouhani B 1993 Acoustic band structure of periodic elastic composites. *Phys. Rev. Lett.* **71**(13) 2022–5.
[3] Akl W and Baz A 2012 Analysis and experimental demonstration of an active acoustic metamaterial cell. *J. Appl. Phys.* **111**(4) 044505.
[4] Akl W and Baz A 2011 Stability analysis of active acoustic metamaterial with programmable bulk modulus. *J. Smart Mater. Struct.* **20**(12) 125010.
[5] Chen X, Xu X, Ai S, Chen H S, Pei Y and Zhou X 2014 Active acoustic metamaterials with tunable effective mass density by gradient magnetic fields. *Appl. Phys. Letters* **105** 071913 https://doi.org/10.1063/1.4893921
[6] Chung D Y and Isler W E 1978 Ultrasonic velocity anisotropy in ferrofluids under the influence of a magnetic feld. *J. Appl. Phys.* **49**(3) 1809–11.
[7] Wright D W and Cobbold R S C 2008 Acoustic wave transmission in time-varying phononic crystals. *Canadian Acoustics* **36**(3) 44-5.
[8] Berenger J-P 1994 A perfectly matched layer for the absorption of electromagnetic waves. *J. Comput. Phys.* **114**(2) 185–200.
[9] Sakamoto S 2008 FDTD calculation of linear acoustic phenomena and its application to architectural acoustics and environmental noise prediction. In: Akiyama I. (eds) *Acoustical Imaging* **29** 459-66.

Composite nanoparticles: applications, creation mechanism, properties

A V Nomoev[1,2,*], V V Syzrantsev[1], N V Yumozhapova[2], E Ch Khartaeva[1], N A Torkhov[3] and K V Zobov[4]

[1]Institute of Physical Materials Science SB RAS, 670031, Ulan-Ude, Sakhyanovoy st., 6, Russia;
[2]Buryat State University, 67000 Ulan-Ude, Smolin st. 24a, Russia;
[3]Tomsk State University, 634050, Tomsk, Lenina av. 30, Russia;
[4]Institute of theoretical and applied mechanics SB RAS, 630090, Novosibirsk, Institutskaya st., 4, Russia.
*e-mail: nomoevav@mail.ru

Annotation This article presents data on the use of Janus-like and core-shell composite nanoparticles consisting of two substances—metal and semiconductor—in nanophotonics and spectroscopy. Investigations were carried out of the mechanism of formation of composite nanoparticles, based on the principle of minimum surface energy of the system and the results of molecular dynamics modelling. Integral representation was employed to analyse size distribution function of nanoparticles, and it was concluded that coagulation predominates during particle formation.
Keywords: composite nanoparticle, surface energy, core-shell particle, Janus nanoparticle, size distribution function of nanoparticles, nanophotonics, high-index dielectric nanoparticles

Introduction

Composite metal and dielectric/semiconductor nanoparticles are promising as the elemental base of nanophotonics, due to the combination of unique properties of both plasmon resonance metal particles and polariton resonant dielectric/semiconductor particles with a high refractive index (high index). In dielectric nanophotonics, the property of high-index dielectric/semiconductor nanoparticles is used to interact intensively with the magnetic component of an electromagnetic wave in the visible spectral region. The spectral position of the corresponding magnetic Mi-resonance can be varied by changing the radius of the spherical nanoparticle, which allows the position to be changed in a wide spectral range, in contrast to plasmon nanoparticles, in which the magnetic response can be obtained only in complex nanostructures and is limited by the IR region for high Joule losses in the optical range. Composite nanoantennas, which combine high-index nanoparticles with plasmon elements, exhibit both electrical and magnetic resonance, suitable for various advanced applications as basic elements for ultrafast optical switching and nonlinear optics [1].

Silicon is used in dielectric nanophotonics due to its high refractive index, about 3 in the visible spectral region, as well as its biocompatibility and biodegradability. A high refractive index leads to local amplification of the electric field near the nanoparticles on nanoantennas created from silicon. Photochemical synthesis of Si nanoparticles coated with Au and Ag along with pure Au and Ag nanoparticles was carried out recently, using IR nanosecond laser treatment of a bulk silicon wafer in aqueous solutions of hydrochloric acid ($HAuCl_4$) and silver nitrate ($AgNO_3$) [2]. Silver is used because of their outstanding plasmonic properties, and, in combination with high-index dielectric nanoparticles of Si, is important as potential building blocks for nonlinear nanophotonics and spectroscopy.

The resonance properties of nanoparticles depend on their shape and structure. The magnitude of the Schottky electric potential at the interface between particles is also determined by the metal / semiconductor interface. Therefore, the synthesis of composite nanoparticles and the mechanisms of their formation are especially relevant at the present time. Composite nanoparticles are used in catalysis to protect particles from oxidation, and in biomedical applications. For example, the antimicrobial effect of Cu/Zn bimetallic nanoparticles is a combination of factors such as a change in the permeability of the bacterial membrane due to zinc ions and the oxidation process caused by copper ions; Fe-Ag particles form a galvanic pair which contribute to the dissolution of nanoparticles, the formation of ions, and enhanced antitumor properties [3]. Under the action of a relativistic electron beam, we obtained the following metal/semiconductor composite nanoparticles with a dielectric silicon part: core shell Cu@SiO_2, Ag@Si, Janus-like $TaSi_2$/Si [4-6]. The authors have presented some electrophysical properties and applications of these composite nanoparticles in the following articles [7-9]. We propose, below, some mechanisms for the formation of composite nanoparticles created by a one-stage gas-phase synthesis method based on two approaches: energy, and molecular dynamics.

Results and discussion

The mechanism for creating composite core/ shell nanoparticles based on the principle of minimum surface energy.

It is known that redistribution occurs at the interface between two liquid media—particles with a lower surface energy diffuse to the interface. In other words, the addition of a small amount of a substance with low surface energy leads to a significant decrease in the surface energy of a two-component liquid, in contrast to the small change in its surface energy following the addition of a small amount of liquid with a large value of surface energy. This is due to the fact that a substance with a small value of surface energy reduces the surface energy of a two-component liquid as a result of segregation to its surface.

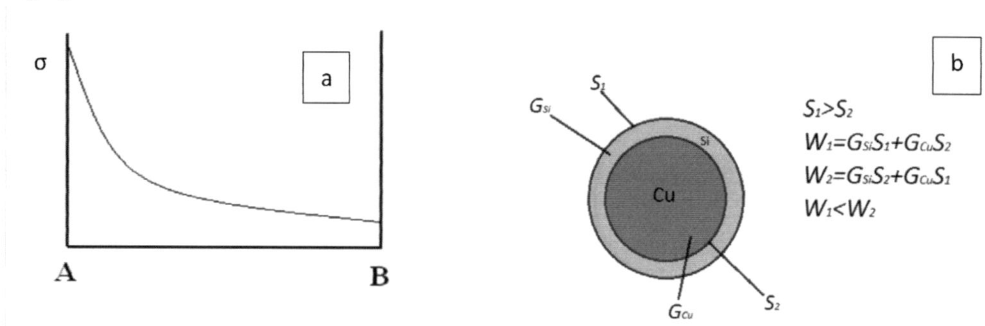

Fig. 1. a - Dependence of the surface energy of a two-component substance on the content of its components; b - surface energy of the core-shell particle.

These substances tend to concentrate in the surface layer, reducing the surface tension of the bicomponent substance. The qualitative behaviour of changes in surface tension depending on the relative content for two-component liquids is shown in Fig. 1a. The surface energy W1 of the composite core-shell particle composed of the sphere with surface area S1 and surface energy G1 and the sphere with the surface area S2 and the surface energy G2 is W = G1S1 + G2S2. In the case when G1> G2 and S1> S2, the surface energy W1 is less than W2 (Fig. 1b). Figure 2 shows a Cu@Si nanoparticle and calculation of the surface energies of the particle. The condition W1 <W2 creates conditions for the segregation of silicon to the surface of the particle, with the formation of a copper core.

Molecular dynamics simulation

Modelling of the process of formation of composite nanoclusters [8] is consistent with experiments on the production of core-shell nanoparticles. The estimated time, based on the results of molecular dynamics modelling, for phase separation does not exceed 1 μs, and the diffusion rate increases significantly with increasing temperature. In the electron beam method used in this work, wherein evaporation by a relativistic electron beam is followed by condensation of vapours in an inert gas stream, high temperatures are reached; therefore, the diffusion rate is high enough for phase separation and there will be enough time for the process of separation and formation of a core-shell particle. The creation of a Cu@Si core-shell nanoparticle from a liquid fcc phase of randomly distributed silicon and copper atoms in a lattice with a 10% silicon content occurs at a cooling rate of 1 K / ps. The formation of Janus-like nanoparticles in the initial configuration in the form of two liquid contacting nanoclusters with a high silicon content of 50% occurs at high cooling rates compared to core-shell nanoparticles. Cu@Si forms a Janus-like nanocluster with a decrease in the cooling rate. The structure and internal structure of the Cu@Si composite cluster depend on the type of initial objects, the concentration of silicon in the alloy, and the crystallisation rate.

Analysis of the size distribution function of composite nanoparticles

Information on the particle formation mechanism is provided by the particles' size distribution. The particle size distribution function was determined from their images obtained using a transmission electron microscope. The number of particles in the statistical sample ranged from 300 to 500. The data obtained for the core-shell particles Cu@SiO_2 are well described by the lognormal distribution (Fig.2), which is observed when the coagulation process prevails at all stages of particle growth at their condensation in the gas phase [1]. It is known that the coagulation process of particle growth is well described by the Smoluchowski kinetic equation [10].

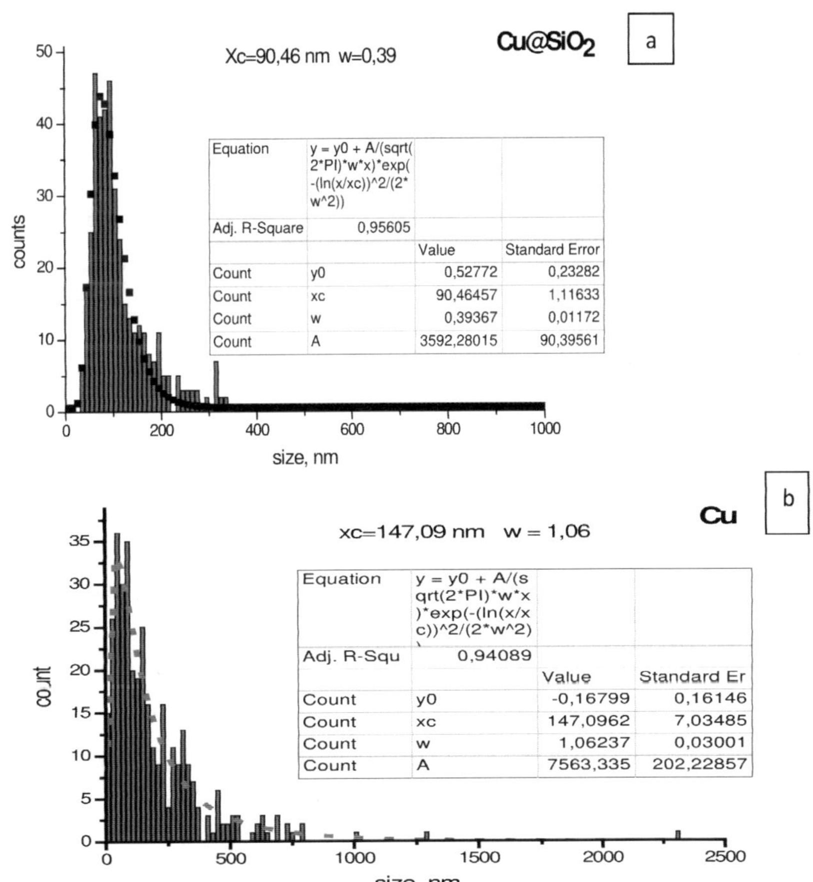

IOP Conf. Series: Materials Science and Engineering **704** (2019) 012018 doi:10.1088/1757-899X/704/1/012018

Fig. 2. Histogram of particle size distribution: a - Cu @ SiO_2, b – Cu

The size distribution function of copper particles is visually close to the lognormal distribution (Fig. 2a), although its deviation from the lognormal will be shown below. As a rule, to estimate the propinquity of the distribution to the lognormal one, the integral probability distribution function on the particle diameter in semilogarithmic coordinates is used. The dependence obtained should be linear. More informative is a comparison of the experimentally obtained distribution with a true lognormal distribution. In this case, the integral probability distribution function in semilogarithmic coordinates is also constructed, and we determine the smallest and largest probabilities where the dependence is linear. The histogram of the experimental values is approximated by the lognormal distribution, and the mean value and the logarithm of the standard deviation are determined. Further, this distribution is constructed in semi-logarithmic coordinates, the probability interval where this dependence is linear is determined, and the obtained interval is compared with the experimental one.

Thus, a large fraction of large sized particles are present in Cu nanopowder as compared to Cu@SiO_2 nanopowder. From these data it follows that the same formation mechanisms are characteristic of the first stages of the formation of Cu@SiO_2 and Cu nanoparticles, whereas a difference is observed at the final stages. The first stage, by the coagulation mechanism, obviously occurs before the formation of the SiO_2 shell and its hardening.

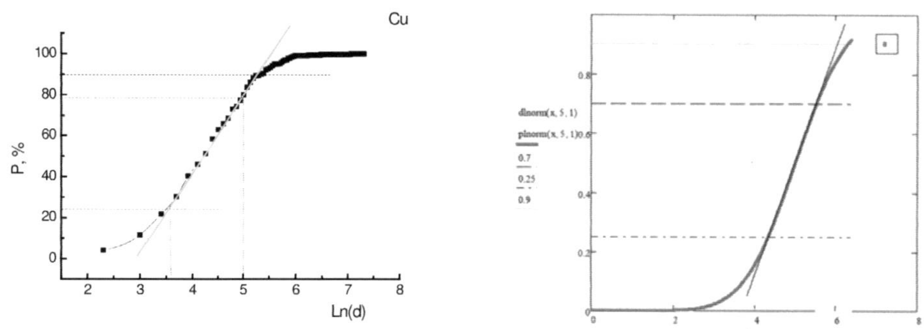

a) b)

Fig. 3. a - integral probability distribution function depending on the particle size Cu in semilogarithmic coordinates, b - true integral probability distribution function depending on the particle size in semilogarithmic coordinates for an average particle size of 147 nm.

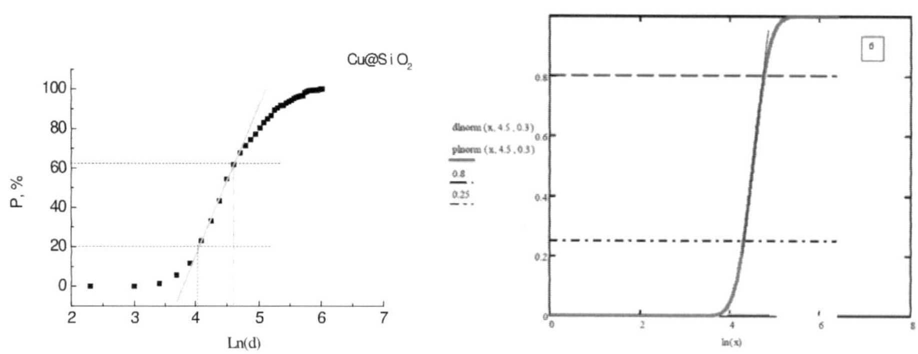

a) b)

Fig. 4. a - integral probability distribution function depending on the particle size Cu@SiO2 in semilogarithmic coordinates, b - true integral probability distribution function depending on the particle size in semilogarithmic coordinates for an average particle size of 90.5 nm.

At the second stage, the coagulation of copper nanoparticles under the shell does not occur due to hardening of the silica shell. This leads to a deviation from the lognormal law when large particles are "not enough" in the distribution, starting with Ln (d) = 4.6 (Fig. 4a). This process most probably occurs at temperatures close to the crystallisation temperature of silicon dioxide, 1400^0C. In the case of copper nanoparticles, the situation is the opposite—in the distribution, there is an "excess" of large particles, starting with Ln (d) = 5, which is most likely due to the manifestation of the liquid-drop mechanism of their formation.

Obviously, one can influence the final particle sizes by choosing a shell material with different condensation temperatures. The arithmetic average size of copper nanoparticles in the shell (90.5 nm) is significantly smaller than the size of pure copper nanoparticles without a shell (190 nm).

For example, if we take a substance for a shell with a high condensation temperature (titanium, tungsten, etc.), the coagulation process will be limited at high temperatures, which will lead to a small particle size. It should be added that the substance of the nucleus should be selected taking into account its surface tension, intermolecular interaction, and the formation energies of chemical reactions with the shell material.

Conclusions

As follows from a review of the published data, Janus-like and the core-shell composite nanoparticles, consisting of a metal and a semiconductor with the dielectric constant of more than 3, are promising for nanophotonics, because of the effects in the particles: plasmon and polariton.

It is shown that the main driving force for the formation of core-shell nanoparticles is the difference in the surface energies of the substances that make up this particle. Addition of a small amount of a substance with a small value of surface energy into a substance with a large value of surface energy leads to a sharp decrease in the total surface energy of the system. Moreover, for nanoscale, due to the high Laplace pressure, and consequently large values of surface energy, this mechanism is predominant.

Based on the experimental data of the authors, on obtaining particles under the action of a relativistic electron beam using the integral size distribution function, it was found that for large particle sizes deviations from the lognormal law, described by the Smoluchowski equation, are observed. The reason for this is the formation of a shell of silicon oxide, as well as the manifestation of a liquid-droplet mechanism.

This work was supported by the Russian Scientific Foundation [project number 18-79-10143].

This experiment was carried out using the equipment of the centre for collective use "Scientific instruments" of the Buryat State University.

References

1. Baranov D G, Zuev D A, Lepeshov S I, Krasnok A E , Evlyukhin A B, Kotov O V and Chichkov B N *Optica* 2017 **4**(7) 814-25
2. M.I. Lerner, A.V. Pervikov, E.A. Glazkova, Svarovskaya N V , Lozhkomoev A S and Psakhie SG 2016 *Powder Technology* **288** 371–8.
3. I.N. Saraeva, N. Van Luongac, S. I. Kudryashov, Rudenko A A, Khmelnitskiy R A, Shakhmin A L, Kharin A Yu, Ionin A A, Zayarny D A, Tung D H, Duong P V and Minh P H 2018 *Journal of Photochemistry and Photobiology A: Chemistry* **360** (1) 125-31.
4. Nomoev A V, Bardakhanov S P, Schreiber M, Bazarova D Z, Baldanov B B and Romanov N A 2015 *Nanomaterials* **5**(1) 26-35.
5. Nomoev A V, Bardakhanov S P, Schreiber M, Bazarova D G, Romanov, N A, Baldanov B B, Radnaev B R and Syzrantsev V V 2015 *Beilstein Journal of Nanotechnology* **6**(1) 874-80.
6. Nomoev A V, Radnaev A R and Kalashnikov S V 2016 *Chemical Physics Letters* **651** 274-7.
7. Kalashnikov S V, Romanov N A and Nomoev A V 2016 *Journal of Applied Physics.* **119**(9) 094304.

8. Nomoev A V, Torkhov N A, Khartaeva E Ch, Syzrantsev V V, Yumozhapova N V, Tsyrenova M A and Mankhirov V N *Chemical Physics Letters* 2019 **720** 113-8.
9. Yumozhapova N V, Nomoev A V and Gafner Y Y 2018 *Solid State Phenomena* **271** 47-50.
10. Smoluchowsky M 1916 *Phys. Zeits.* **17** 557-85.

IOP Conf. Series: Materials Science and Engineering **704** (2019) 012019 doi:10.1088/1757-899X/704/1/012019

Research of silicon and silver nanoparticles obtained by laser ablation in liquid

E I Gyrylov

Institute of Physical Materials Science, Siberian Branch of the Russian Academy of Sciences, 6 Sakhyanovoy Street, Ulan-Ude 670047, Russian Federation

E-mail: evivgyr.bsc@mail.ru

Abstract. Laser ablations of silicon and silver in ethanol were made and the nanoparticles obtained were studied. It was shown the silver nanoparticles had both monocrystal and polycrystalline structures and cubic and hexagonal structures. It was cleared the amorphous silicon nanoparticles included a little quantity of the monocrystalline silicon nanoparticles.

1. Introduction

Laser ablation refers to the method of evaporation of a substance from a surface by a laser pulse. During laser ablation of solids in liquids, the high density of the buffer liquid prevents the laser plume from expanding, the vapor-plasma cloud escaping from the target quickly cools with the formation of small particles, the nanoparticles remain in the liquid volume, forming a colloidal solution.

Pulse laser ablation of substances in a liquid is applicable both for all types of solids [1, 2]. Ablation of solids in a liquid allows obtaining nanosuspensions and eliminates the presence of residual reagents in the final product and additional purification, in contrast to the chemical synthesis of nanoparticles in a liquid.

For a nanosecond laser, most of the laser energy comes to the target after the formation of a vapor layer. Target melting occurs after a boiling crisis (complete coating of the target surface with a vapor film). Ablation of the material following melting occurs in the atmosphere vapor of liquid (which subsequently forms cavitation bubbles above the target surface) [3].

2. Experimental

For ablation we used an ytterbium pulsed fiber laser YLP (Russia). The laser emitted periodic pulses with a wavelength of 1064 nm, pulse duration of 100 ns, pulse energy of 1 mJ and a peak pulse power of up to 10 kW. [4]

Ingots of Ag and Si were used as targets and an ethanol solution was used as dispersion medium. The silver has low chemical activity, therefore, pure metal particles are formed in ethanol.

The laser radiation was focused by a collecting lens onto a target sample, which was placed at the bottom of the liquid container. The sample was scanned at a speed of 0.3-1 mm / s using a motorized biaxial linear translator 8MTF (Lithuania) and a CNC USB TB6560 controller (China). Multiple scanning of a given surface of the sample was carried out within 10-30 minutes. The use of the scanning system made it possible to uniformly evaporate the material from the target, thereby increasing the ablation efficiency during prolonged irradiation and narrowing the size spread of the nanoparticles obtained.

Colloidal solutions were studied using a SALD-7500nano laser particle size analyzer (Japan). At a power density in the pulse of 8 × 10^7 W / cm^2 and a pulse frequency of 40 kHz, the average particle size in the solution was 50 nm for Ag and 14 μm for agglomerates of coagulated Si particles. After processing a colloidal Si solution in an ultrasonic bath with a frequency of 40 kHz and a power of 30 W, the average particle size of Si in the solution was 35 nm.

3. Results and analysis

The colloidal solutions obtained were investigated by transmission electron microscopy (TEM) (Figure 1). It indicates diffraction lines of the crystal structure. The interplanar distances *d* are shown of double light segments. Ag nanoparticles have two structures. The first is a polycrystalline correspond to the cubic structure (interplanar distances of 0.236 nm or 0.205 nm) or the hexagonal structure (interplanar distances of 0.227 nm) (Figure 1a). The second is a monocrystal structure which corresponds to a hexagonal close packed structure (interplanar distance of 0.48 nm) [5] (Figure 1b).

Silicon nanoparticles are mainly amorphous, but there are also monocrystal structures (Figure 1c) with an interplanar distance of 0.19 nm (cubic diamond structure).

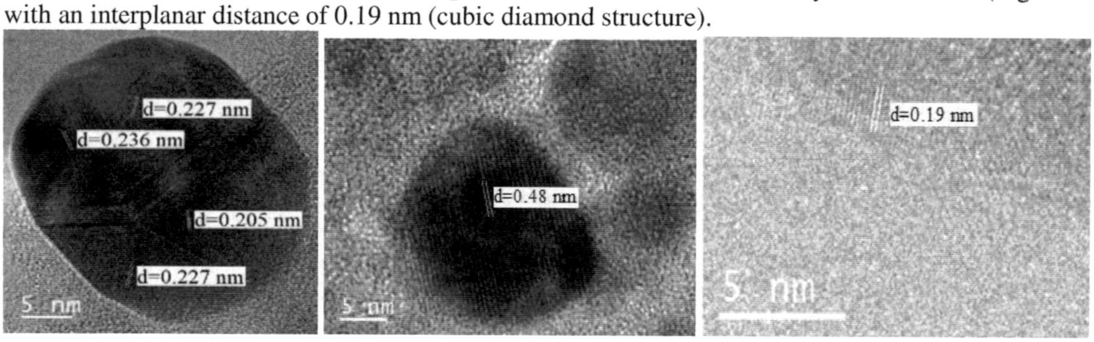

(a) (b) (c)

Figure 1. TEM images of nanoparticles: (a) TEM image of Ag polycrystalline nanoparticle; (b) TEM image of Ag single-crystal nanoparticle; (c) TEM image of Si single-crystal nanoparticle.

TEM image shows that Ag nanoparticles have a shape which is close to spherical in size 24×30 nm^2 (Figure 1a) and 20×22 nm^2 (Figure 1b), and crystalline Si nanoparticles have an oblong shape of 2×8 nm^2 (Figure 1c). X-ray phase analysis confirmed the presence of crystalline silicon. Figure 2 shows X-ray diffraction patterns for silver and silicon nanoparticles. X-ray diffraction pattern (Figure 2a) shows the interplanar spacings of the crystal lattice for Ag nanoparticles to be 2.360 Å, 2.047 Å, 1.445 Å, 1.231 Å and (Figure 2b) for Si: 3.141 Å, 1.922 Å, 1.636 Å, 1.246 Å.

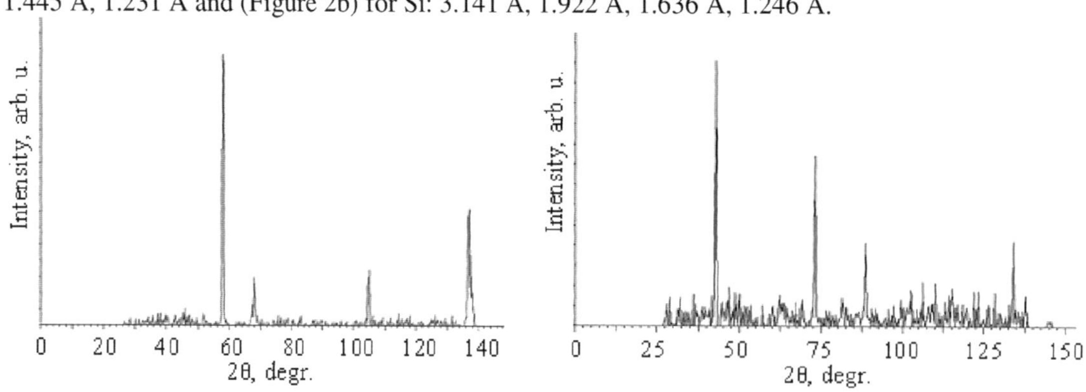

(a) (b)

Figure 2. X-ray diffraction patterns of particles: (a) for Ag; (b) for Si.

4. Conclusions

Thus, colloidal solutions of silicon and silver nanoparticles with shape close to spherical with an average diameter of 35 nm and 50 nm, respectively, were obtained. Silver nanoparticles have both monocrystal and polycrystalline structures and exist in two polymorphic structures, cubic and hexagonal. Along with amorphous silicon nanoparticles, oblong 8 nm monocrystalline silicon nanoparticles are also present.

This work has been performed under the program of fundamental research of state academy of sciences (project 0336-2016-0006).

References

[1] Nomoev A V, Torhov N A, Khartaeva E Ch, Syzrantsev V V, Yumozhapova N V, Tsyrenova M A and Mankhirov V N 2019 *Chem.Phys.Letters* **720** 113–8 https://doi.org/10.1016/j.cplett.2019.02.015

[2] Kazakevich P V, Voronov V V, Simakin A V and Shafeev G A 2004 Formation of copper and brass nanoparticles during laser ablation in a liquid. *Kvantovaya elektronika* **34** (10) 951-956.

[3] Tsuji T, Okazaki Y, Tsuboi Y and Tsuji M 2007 Nanosecond time-resolved observations of laser ablation of silver in water *Japan. J. Appl. Phys.* Part 1. **46** (4A) 1533–5.

[4] Gyrylov E I and Nomoev A V 2018 *Zavodskaya laboratoriya. Diagnostika materialov* **84**(9) 41-5.

[5] Murzakaev A M 2017 Size dependence of the phase composition of silver nanoparticles formed by the electric explosion of a wire. *The Physics of Metals and Metallography* **118** (5) 459-65.

AUTHOR INDEX